# Mastering Mobile Test Automation

Master the full range of mobile automation and testing techniques to develop customized mobile automation solutions

**Feroz Pearl Louis**

**Gaurav Gupta**

[PACKT] open source*

PUBLISHING

community experience distilled

BIRMINGHAM - MUMBAI

# Mastering Mobile Test Automation

First published: April 2015

Production reference: 1270415

Published by Packt Publishing Ltd.
Livery Place
35 Livery Street
Birmingham B3 2PB, UK.

ISBN 978-1-78217-542-1

www.packtpub.com

# Credits

# About the Authors

**Feroz Pearl Louis** is an automation architect with over 10 years of extensive experience working with automation tools, such as HP UFT(formerly QTP), Selenium, SOAtest, SoapUI Pro, Cucumber, Watir, Ranorex, SeeTest, Perfecto Mobile, and Mobile Labs. His core expertise is in functional and mobile automation with traditional and nontraditional tools and techniques. He is currently working with the Automation Centre of Excellence of a Fortune 500 IT services company. He leads a team of 40 plus people spread across the entire spectrum of automation services, including functional, SOA, mobile, cross-browser and cross-platform automation, and Agile automation with behavior-driven development methodology. With his extensive design and development experience, he engages with various customer teams to design, implement, and deliver solutions that address their specific pain points. He is a prolific contributor to the automation community on various forums, such as LinkedIn, Quora and SQA Forums, and blogs at `https://ferozlouis.blogspot.com`.

**Gaurav Gupta** is a dynamic and young IT professional who is engaged in the testing of web and mobile applications. He has worked with various tools and techniques in the broad field of mobile testing. He has extensive experience in both web and mobile test suite design and optimizes various operating systems. He is a versatile tester and is always keen to learn new technologies to keep himself updated. His passion for work makes him stand out among others. Even at a relatively early stage of his career, he is the published author of the book *Mastering HTML5 Forms*. A graduate in Computer Science, he works with a reputed Fortune 500 IT services company. Gaurav is a native of Chandigarh, India, and currently lives in Pune, India.

# Acknowledgements

First and foremost,I would like to thank my family — especially my mother — the guiding light of my life and my better half, Sadaf, whose constant support and encouragement made it possible for me to burn that midnight oil and brought me back from the verge of giving up. I would also like to thank my son, Haaris, who understood myneed for long nights bythe computer.I would like to specially thank my sister,Dinshaw, who has always been a constant support - **Feroz Pearl Louis**

First of all, I would like to thank the almighty and my family, who have always guided me to walk on the right path oflife. My heartfelt gratitude and indebtedness goes to all those people in my life who gave me constructive criticism, as it contributed directly or indirectly in a significant way towards firing up my zeal to achieve my goals - **Gaurav Gupta**

We also gratefully acknowledge, with a deep sense of gratitude and most sincere appreciation, the timely support provided by Ashwarya Singh during the process of writing this book.

We wish to extend our sincere gratitude to the team from Packt Publishing and the technical reviewers for their valuable suggestions, which proved extremely helpful in making this a better book for you.

Our special thanks to our mentors, colleagues,and friends for sharing their mobile testing experiences, which have proved very valuable in making this book better oriented toward the real-world challenges faced in mobile testing projects.

# About the Reviewers

**Karl Chavarria**, LFCE, is a software engineer in test currently working for LivingSocial and is also an avid technology hobbyist. For the past 7 years, Karl has lived in the Portland metro area, where he has worked at various tech start-ups and studied Computer Science at Portland State University. Karl started his career in Linux system administration, but soon steered toward quality assurance after seeing just how much it incorporated one of his favorite tasks: automation. When he's not increasing the efficiency of his Ruby code, Karl can be found working on programming-side projects, playing music arcade games, and enjoying all of the delicious food that the Pacific Northwest has to offer.

**Arik Lewin** has many years of experience in the software development of changing environments and technologies with server-side and client-side orientation. He has participated in many successful start-ups and enterprise large-scale projects as a key player, expert developer, innovator, and architect. Over the years, he acquired unique skills from the web tier by creating mobile and desktop applications. He has worked with frameworks from Java to, the most popular, JavaScript.

Over the past few years, Arik has joined HP Mercury, participated in making test automation products, and initiated his own open source project automation framework for mobile web named CATJS. He attended many related events as a speaker and participated in promoting the new agile methodology and approach for mobile web testing.

Arik helped a lot of development groups to make versatile challenges and still shares his vast knowledge with others.

**Luis Lu** is a testing architect and inventor at SAP Labs, China. He has solid technology experience in quality assurance, web/mobile test automation, performance, security, continuous integration, and so on. He is an open source fan and likes to play around with new technology. You can reach him directly at his GitHub website `https://github.com/luisxiaomai` or e-mail him at `luzhenyuhnr@126.com`.

# www.PacktPub.com

## Support files, eBooks, discount offers, and more

For support files and downloads related to your book, please visit www.PacktPub.com.

Did you know that Packt offers eBook versions of every book published, with PDF and ePub files available? You can upgrade to the eBook version at www.PacktPub.com and as a print book customer, you are entitled to a discount on the eBook copy. Get in touch with us at service@packtpub.com for more details.

At www.PacktPub.com, you can also read a collection of free technical articles, sign up for a range of free newsletters and receive exclusive discounts and offers on Packt books and eBooks.

https://www2.packtpub.com/books/subscription/packtlib

Do you need instant solutions to your IT questions? PacktLib is Packt's online digital book library. Here, you can search, access, and read Packt's entire library of books.

## Why subscribe?

- Fully searchable across every book published by Packt
- Copy and paste, print, and bookmark content
- On demand and accessible via a web browser

## Free access for Packt account holders

If you have an account with Packt at www.PacktPub.com, you can use this to access PacktLib today and view 9 entirely free books. Simply use your login credentials for immediate access.

# Table of Contents

# Preface

This book will guide you to develop an in-depth understanding of mobile automation concepts and techniques to design, implement, and deliver customized mobile automation solutions with real, emulated devices and cloud-based mobile testing tools, such as Perfecto Mobile, and Mobile Labs, integrated with traditional tools, such as UFT and Selenium, and other leading tools in the market, such as ExperiTest SeeTest.

This book is a comprehensive reference guide to every aspect of mobile test automation and can be used in any organization that is engaged in test automation and /or takes up mobile test automation.

# What this book covers

*Chapter 1, Ensuring Five-star Rating in the MarketPlace,* explains how mobile test automation is different from traditional functional automation and what the typical challenges are that you have to overcome. This chapter provides you with a brief overview of the different techniques used for mobile automation. These techniques are elaborated in later sections of the chapter. This section explains why and how there are different techniques.

*Chapter 2, Designing Mobile Automation Frameworks,* explains the various types of automation frameworks with respect to mobile automation, such as functional decomposition, keyword-driven, data-driven, and hybrid, with a focus on mobile-specific differences. This has real examples of working code for various tools.

*Chapter 3, User Agent – automating Mobile Applications with Browsers,* provides you with the opportunity to learn the use of the user-agent technique, along with the code snippets and lab setup for this technique. We will also delve into the advantages and disadvantages of this technique.

*Chapter 4, Emulators and Simulators – the Automation of Emulated Devices*, explains the use of emulators and simulators for mobile test automation, along with the code snippets and lab setup for this technique. We will also delve into the advantages and disadvantages of this technique.

*Chapter 5, Automating Physical Devices*, showcases the techniques required to automate a mobile application with physically present real mobile devices.

*Chapter 6, Automating on Cloud*, provides you with the opportunity to learn the techniques required to automate a mobile application with cloud tools such as Perfecto Mobile and Mobile Labs.

*Chapter 7, Optimizing Test Strategy and Estimation*, deals with the selection of the most suitable strategy to focus on maximizing the return on investment for automation and the ways to avoid any potential pitfalls. We will also look at the various technical aspects that are usually overlooked and only become apparent at the end of a delivery cycle. This chapter deals with how to estimate your mobile automation effort.

*Chapter 8, Delivering Customer Delight*, sums up all the lessons learnt in the previous chapters and covers the test strategy planning, project effort optimization, ways to maintain the automation suite, resource identification, and the management of hardware and software.

# What you need for this book

A familiarity with functional automation testing would help, but it is not a prerequisite as all concepts are explained in a detailed manner in this book. To work with the examples practically, you need to use any given tool for mobile test automation such as:

- Selenium (WebDriver 2.0 with Android and iOS Driver)
- Appium
- HP UFT 12.00 or a later version
- Perfecto Mobile
- ExperiTest SeeTest

# Who this book is for

The core audiences are mobile application developers, QAs, project managers, and automation and manual testers who are beginning to learn or are working on any mobile testing assignment.

# Conventions

In this book, you will find a number of text styles that distinguish between different kinds of information. Here are some examples of these styles and an explanation of their meaning.

Code words in text, database table names, folder names, filenames, file extensions, pathnames, dummy URLs, user input, and Twitter handles are shown as follows: "The keywords array contains the list of all keywords for a particular test case denoted by a row in the `BusinessFlow` excel file."

A block of code is set as follows:

```
'Test Script 1: To validate that application Login is allowed with
valid credentials
'Open EMail
fn_gotoTestingEMailAccount();
'Enter the details
fn_EnterDetails();
'Submit
fn_SubmitToCreate();
```

**New terms** and **important words** are shown in bold. Words that you see on the screen, for example, in menus or dialog boxes, appear in the text like this: "After installation of Firefox User Agent add-on from the add-on market, it is accessible from **Tools** menu."

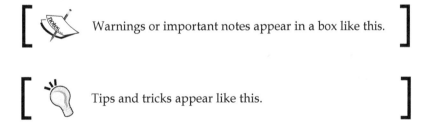

Warnings or important notes appear in a box like this.

Tips and tricks appear like this.

# Reader feedback

Feedback from our readers is always welcome. Let us know what you think about this book—what you liked or disliked. Reader feedback is important for us as it helps us develop titles that you will really get the most out of.

To send us general feedback, simply e-mail feedback@packtpub.com, and mention the book's title in the subject of your message.

If there is a topic that you have expertise in and you are interested in either writing or contributing to a book, see our author guide at www.packtpub.com/authors.

# Customer support

Now that you are the proud owner of a Packt book, we have a number of things to help you to get the most from your purchase.

# Downloading the example code

You can download the example code files from your account at http://www. packtpub.com for all the Packt Publishing books you have purchased. If you purchased this book elsewhere, you can visit http://www.packtpub.com/support and register to have the files e-mailed directly to you.

# Errata

Although we have taken every care to ensure the accuracy of our content, mistakes do happen. If you find a mistake in one of our books—maybe a mistake in the text or the code—we would be grateful if you could report this to us. By doing so, you can save other readers from frustration and help us improve subsequent versions of this book. If you find any errata, please report them by visiting http://www.packtpub. com/submit-errata, selecting your book, clicking on the **Errata Submission Form** link, and entering the details of your errata. Once your errata are verified, your submission will be accepted and the errata will be uploaded to our website or added to any list of existing errata under the Errata section of that title.

To view the previously submitted errata, go to https://www.packtpub.com/books/content/support and enter the name of the book in the search field. The required information will appear under the **Errata** section.

# Piracy

Piracy of copyrighted material on the Internet is an ongoing problem across all media. At Packt, we take the protection of our copyright and licenses very seriously. If you come across any illegal copies of our works in any form on the Internet, please provide us with the location address or website name immediately so that we can pursue a remedy.

Please contact us at `copyright@packtpub.com` with a link to the suspected pirated material.

We appreciate your help in protecting our authors and our ability to bring you valuable content.

# Questions

If you have a problem with any aspect of this book, you can contact us at `questions@packtpub.com`, and we will do our best to address the problem.

# 1
# Ensuring Five-star Rating in the MarketPlace

The *star* rating system on mobile marketplaces, such as Google Play and Application Store, is a source of positive as well as negative feedback for the applications deployed by any organization. This system is used to measure various aspects of the application, such as like functionality, usability, and is a way to quantify the all-elusive measurement-defying factor that organizations yearn to measure called "user experience", besides the obvious ones, such as the appeal and aesthetics of an application's **graphical user interface** (**GUI**). If an organization does not spend time in testing the functionality adequately, then it may suffer the consequences and lose the market share to competitors. The challenge to enable different channels such as web applications through mobile browsers, as well as providing different native or hybrid applications to service the customers as per their preferences, often leads to a situation where organizations have to develop both a web version and a hybrid version of the application.

At any given point of time, it is almost impossible to test an application completely, and to cover various permutations and combinations of operating systems, their versions, device manufacturers, device specifications with various screen sizes, and application types, with solely employed manual testing techniques.

This is where automation comes to the rescue. However, mobile automation in itself is very complex because of the previously explained fragmentation issue. In this chapter, you will learn how not to fall into the trap of using different tools, frameworks, and techniques to address these differences.

In this chapter, we will cover the following topics:

- Introduction to mobile test automation
- Types of mobile application packages
- Mobile test automation overview
- Some common factors to be considered during mobile testing, including Interrupt testing, form factor testing, layout testing, and more
- Overview of different types of mobile automation testing approaches
- Selection of the best mobile testing approach depending on the project
- Troubleshooting and best practices

# Introduction to mobile test automation

Before we start learning about mobile test automation, let's understand what functional test automation is.

Test automation has always been a fundamental part of the software testing lifecycle for any project. Organizations invariably look to automate the repetitive testing actions in order to utilize the manual effort thus saved for more dynamic and productive tasks. Use of automation tools also allows utilization of system idle time more effectively. To address these needs, there are a plethora of tools available in the market along with various frameworks and implementation techniques. There are both open source and licensed tools available in the market. Tools such as HP's **Unified Functional Testing (UFT)**, formerly known as **QuickTest Professional (QTP)**, **TestComplete, Selenium, eggPlant, Ranorex, SilkTest, IBM Functional tester**, and numerous others, provide various capabilities for functional automation.

However, almost all of these tools are designed to support only a single operating system (predominantly Windows—owing to its popularity and the coverage it enjoys across industry verticals), although a few provide support for other lesser-used operating systems, such as Unix, Linux, Sun Solaris, and Apple Macintosh.

As far as functional automation is concerned, you don't need to even consider the implications of supporting multiple operating systems in most cases. With Windows as the only operating system that is supported, there aren't any considerations for different operating systems. If the application is a web application, then there may be a need to do cross-browser testing, that is, testing automation on various browser types (Chrome, Firefox, and Safari besides Internet Explorer) and their respective versions.

Also, as far as functional automation is considered, there is a very clear demarcation between nonfunctional and functional requirements. So, an automated solution for functional testing is not required to consider factors such as how others processes running on the machine would impact it, or any of the hardware aspects, such as the screen resolution of monitors and the make of the machines (IBM, Lenovo, and others).

When it comes to mobile automation, there is an impact on the test suite design due to various other aspects, such as operating systems (Android, iOS, Blackberry, Windows) on which the application is supposed to be accessed, the mode of access (Wi-Fi, 3G, LTE, and so on), the form factor of the devices (tablets, phones, phablets, and so on), and the behavior of the application in various orientation modes (portrait, landscape, and so on).

So, apart from normal automation challenges, a robust mobile automation suite should be able to address all these challenges in a reliable way. In the coming chapters, we will learn about how to create such frameworks, but before that, we need to have a thorough understanding of the challenges.

Fragmentation of the mobile ecosystem is an aspect that compounds this manifold problem. An application should be able to service different operating systems and their flavors provided by **original equipment manufacturers (OEMs)**, such as Apple with iOS, Google's Android with Samsung, HTC, Xiaomi, and numerous others, Windows with Nokia and HTC, and even Blackberry and other lesser-used operating systems and devices. Add to this the complexity of dealing with various form factors, such as phones, tablets, phablets, and their various hybrids.

The following figure is a visualization of the Android market fragmentation over various equipment manufacturers, form factors, and OS versions:

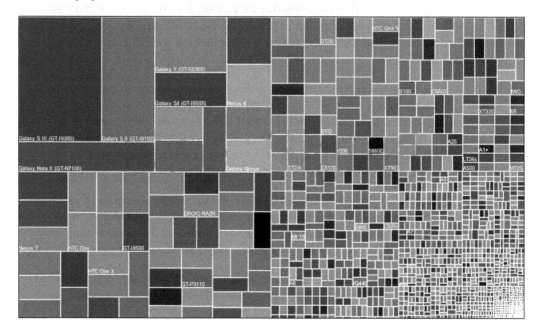

As we know, test automation is the use of software to automate and control the setting up of test preconditions, execution of tests, test control, and test reporting functions with minimum, or ideally zero, user intervention. Automating the testing for any mobile application is the best way to ensure quality, and to achieve the quick and precise results that are needed to accommodate fast development cycles.

Organizations look toward functional test automation primarily to reduce the total cost of ownership over a period of time, and to ensure the quality of the product or application being developed. These advantages are compounded many times for mobile test automation and hence it provides the same advantages, but to a much greater degree.

The following are the various advantages of mobile test automation for any project:

- **Improved testing efficiency**: The same scripts can be used to run uniform validations across different devices, operating systems, and application types (of the same application), thereby reducing the test execution effort considerably. This also means that the **return on investment** (**RoI**), which typically takes about 3-5 cycles of executing the conventional functional automation to achieve breakeven, is viable in most cases within the first release itself, as mobile testing is typically repeated on many devices. So, in this case, fragmentation acts as a positive factor if the automation is employed properly, whereas, with pure manual testing, it greatly increases the costs.

- **Consistent and repeatable testing process**: Human beings tend to get bored with repetitive tasks and this makes such a test prone to errors. Due to the effect of fragmentation in the mobile world, the same application functionality needs to be validated across various combinations of operating systems, application types, device manufacturers, network conditions, and many more. Hence, the use of automation, which is basically a program, ensures that the same scripts run without any modifications every time.

- **Improved regression testing coverage**: The use of automation scripts allows the regression tests to be iterated over multiple combinations of test data. Such data-driven scripts allow the same flow to be validated against different test data combinations. For example, if an application allows users to search for the nearest ATMs in a given area, basically, the same flow would need to be tested with various zip codes as inputs. Hence, the use of automated scripts would instantly allow the test coverage to be increased dramatically.

- **More tests can be run in less time**: Since automated scripts can be run in parallel over various devices, the same amount of testing can be compacted inside a much smaller time window in comparison to the manually executed functional testing. With the use of automation scripts that include device setups as preconditions, the execution window can be exponentially reduced, which otherwise would take a manual tester considerable time to complete.

- **24/7 operation**: Although any functional automation suite can lead to better resource utilization in terms of executing more number of scripts in lesser time, with respect to mobile automation, the resources are often expensive mobile devices. If functional testing is done manually, then more of the same devices need to be procured to allow manual testers to carry out tests, and especially, more so in the case of geographically distributed testing teams. Mobile automation scripts, on the other hand, can be triggered remotely and can run unattended, reducing the overall cost of ownership and allowing 24/7 utilization of devices and tools.

- **Human resources are free to perform advanced manual tests**: Having automation scripts perform repetitive regression testing tasks frees up the bandwidth of manual testing teams for exploratory tests that are expensive to automate and cumbersome to manage. Hence, the use of automation leads to a balanced approach, where testers can perform more meaningful work and thereby improve the quality of delivered applications. In mobiles, since regression is more repetitive on account of the fragmentation problem, the amount of effort saved is manifold, and hence, testers can generally focus on testing aspects such as **user interface (UI)** testing and user experience testing.

- **Simple reproduction of found defects**: Since automation scripts can be executed multiple times on demand and are usually accompanied with reports and screenshots, defect triangulation is easy and is just a matter of re-execution of automation scripts. With pure manual testing, a tester would have to spend effort on manually recreating the defect, capturing all the required details, and then reporting it for defect tracking. With mobile automation, the same flow can be triggered multiple times on a multitude of devices hence, the same defect can be replicated and isolated if it occurs only on a specific set of devices.

- **Accurate and realistic real-life mobile scenarios**: Since a mobile requires tests to be specifically designed for variable network conditions and other considerations, such as device screen sizes, orientation, and more, which are difficult to recreate accurately with pure manual testing effort, automation scripts can be developed that accurately to recreate these real-world scenarios in a reliable way. These types of tests are mainly not required to be developed for functional automation suites, and hence, this is one of the major differences.

For the most realistic results, conventional wisdom is to test automation on actual devices—without optical recognition, emulation, jailbreaking, or tethering. It is impractical to try to automate everything, especially for mobile devices. However, leveraging **commercial off-the-shelf (COTS)** tools can vastly reduce the cost of automation and thereby enhance the benefits of the automation process.

In the following section, we will discuss in detail the challenges that make mobile automation vastly different from conventional functional automation.

The following are some of the issues that make the effective testing automation of mobile applications challenging:

- **Restricted access to native methods to enable automation tools**: Traditional functional automation tools utilize native operating system methods to emulate user interactions. This is comparatively easy to do as the operating system allows access. However, the same level of access is not available with a mobile operating system. Also, inter-application interactions are restricted in a mobile operating system and each application is treated as an individual thread. This is normally only allowed when a phone is rooted or when the application under test is modified to allow instrumentation access. So, using other software (the test automation tool) to control user inputs in a mobile application is much more difficult to achieve and consequently slower or more error prone.

  For example, if an Android application under test makes a call to the photo gallery, then the automated test would not be able to continue because a new application comes to the foreground.

- **Lack of prediction techniques for UI synchronization in a Mobile environment**: In addition to the restricted access mentioned in the previous point, mobile application user interface response times are dependent on many variables, such as connection speed and device configuration other than the server response times. Hence, it is much harder to predict the synchronization response in a mobile application. Due to this automation of mobile, the application is more prone to be unstable unless hardcoded wait times are included in the automation scripts.

- **Handling location-specific changes in the application behavior**: Many mobile applications are designed to interact with the user location, and behave differently as per the change in GPS coordinates. Since network strengths cannot be controlled externally, it is very difficult to predict the application behavior and to replicate the preconditions of a network strength-specific use case through the use of automation. So, this is another aspect that every automation solution has to address appropriately. Some automation tools allow the simulation of such network conditions that should be specifically handled while developing the automation suite.

- **Supporting application behavior changes for varied form factors**: As explained earlier, since there are different screen sizes, available for mobile devices, the behavior of the application is often specific to the screen size owing to responsive design techniques that are now quite widely used. Even with the change in the orientation of the devices, application use cases have alternative behavior. For example, an application interface loaded in the portrait mode would appear different, with objects in different locations than they would appear in the landscape mode. Hence, automation solutions would need to factor this in and ensure that such changes are handled in a robust and scalable way.

- **Scripting complexity due to diversity in OS**: Since many applications are developed to support various OSes, especially mobile web applications, it is a key challenge to handle application differences, such as mobile device input methods for various devices, as devices differ in keystrokes, input methods, menu structures, and display properties. With different mobile operating systems in the market, such as Android, iOS, Brew Symbian, Tizen, Windows, and BlackBerry (RIM), each having its own limitations and variations, creation of a single script for every device is a challenge that needs to be adequately tackled in order to make the automation solution more robust, maintainable, and scalable to support newer devices in future.

In subsequent chapters, we will discuss how to effectively manage these using various techniques.

# Mobile application packages

With the advancement in wireless technology, big technology companies, such as Apple, Amazon, Google, and so on, came out with a solution that provides users with a more realistic approach to finding information, making decisions, shopping, and other countless things at their fingertips by developing mobile applications for their products. The main purpose of developing mobile applications was actually to retrieve information using various productivity tools, which includes calculator, e-mail, calendar, contacts, and many more. However, with more demand for and the availability of resources, there was a rapid growth and expansion in other categories, such as mobile games, shopping, GPS and location-based services, banking, order tracking, ticket purchases, and recently, mobile medical applications.

The distribution platforms, such as Apple App Store, Google Play, Windows Phone Store, Nokia Store, and BlackBerry Application World, are operated by the owners of the mobile operating systems, and mobile applications are made available to users by them. We usually hear about the terms such as a native application, hybrid application, or web application, so, did you ever wonder what they are and what is the difference is between them? Moving ahead, we will discuss the different mobile packages available for use and their salient features that make an impact on the selection of a strategy and testing tool for automation.

The different mobile packages available are:

- Native applications
- Web applications
- Hybrid applications

# Native applications

Any mobile application needs to be installed through various distribution systems, such as Application Store and Google Play. Native applications are the applications developed specifically for one platform, such as iOS, Android, Windows, and many more. They can interact and take full advantage of operating system features and other software that is typically installed on that platform. They have the ability to use device-specific hardware and software, such as the GPS, compass, camera, contact book, and so on.

These types of applications can also incorporate gestures such as standard operating system gestures or new application-defined gestures. Native applications have their entire code developed for a particular operating system and hence have no reusability across operating systems. A native application for iOS would thus have its application handles built specifically for Objective-C or Swift and hence would not work on an Android device. If the same application needs to be used across different operating systems, which is a very logical requirement for any successful application, then developers would have to write a whole new repository of code for another mobile operating system.

This makes the application maintenance cumbersome and the uniformity of features is another challenge that becomes difficult to manage. However, having different code bases for different operating systems allows the flexibility to have operating-system-specific customizations that are easy to build and deploy. Also, today, there is a need to follow very strict "look and feel" guidelines for each operating system. Using a native application might be the best way to keep this presentation correct one for each OS.

Also, testing native applications is usually limited to the operating system in question and hence, the fragmentation is usually limited in impact. Only manufactures and operating system versions need to be considered.

# Mobile web applications

A mobile web application is actually not an application but in essence only websites that are accessed via a mobile interface, and it has design features specific to the smaller screen interface and it has user interactions such as swipe, scroll, pinch, and zoom built in. These mobile web applications are accessed via a mobile browser and are typically developed using HTML or HTML5. Users first access them as they would access any web page. They navigate to a special URL and then have the option of installing them on their home screen by creating a bookmark for that page.

So, in many ways, a web application is hard to differentiate from a native application, as in mobile screens, usually there are no visible browser buttons or bars, although it runs in mobile browsers. A user can perform various native application functionalities, such as swiping to move on to new sections of the application.

Most of the native application features are available in the HTML5 web application, for example, they can use the tap-to-call feature, GPS, compass, camera, contact book, and so on. However, there are still some native features that are inaccessible (at least for now) in a browser, such as the push notifications, running an application in the background, accelerometer information (other than detecting landscape or portrait orientations), complex gestures, and more.

While web applications are generally very quick to develop with a lot of ready-to-use libraries and tools, such as **AngularJS**, **Sencha**, and **JQuery,** and also provide a unique code base for all operating systems, there is an added complexity of testing that adds to the fragmentation problem discussed earlier. There is no dearth of good mobile browsers and on a mobile device, there is very limited control that application developers can have, so users are free to use any mobile browser of their choice, such as Chrome, Safari, UC Browser, Opera Mobile, Opera Mini, Firefox, and many more. Consequently, these applications are generally development-light and testing-heavy. Hence, while developing automation scripts, the solution has to consider this impact, and the tool and technique selected should have the facility to run scripts on all these different browsers.

Of course, it could be argued that many applications (native or otherwise) do not take advantage of the extra features provided by native applications. However, if an application really requires native features, you will have to create a native application or, at least, a hybrid application.

# Hybrid applications

Hybrid applications are combinations of both native applications and web applications, and because of that, many people incorrectly call them web applications. Like native applications, they are installed in a device through an Application Store and can take advantage of the many device features available. Just like web applications, hybrid applications are dependent on HTML being rendered in a browser, with the caveat that the browser is embedded within the application. So, for an existing web page, companies build hybrid applications as wrappers without spending significant effort and resources, and they can get their existence known in Application Store and have a star rating! Web applications usually do not have one and hence have this added disadvantage of lacking the automatic publicity that a five-star rating provides in the mobile stores.

Because of cross-platform development and significantly low development costs, hybrid applications are becoming popular, as the same HTML code components are reusable on different mobile operating systems. The other added advantage is that hybrid applications can have the same code base wrapped inside an operating-system-specific shell thereby making it development-light. By removing the problem posed by various device browsers, hybrid applications can be more tightly controlled, making them less prone to fragmentation, at least on the browser side. However, since they are hybrid applications, any automation testing solution should have the ability to test across different operating system and version combinations, with the ability to differentiate between various operating-system-specific functionality differences. Various tools such as **PhoneGap** and **Sencha** allow developers to code and design an application across various platforms just by using the power of HTML.

# Factors to be considered during mobile testing

In many aspects, an approach to perform any type of testing is not so different from mobile automation testing. From methodology and experience, while working with the actual testing tools, what testers have learned in testing can be applied to mobile automation testing.

So, a question might come to your minds that then, where does the difference lie and how should you accommodate these differences? So, following this topic, we will see some of the factors that are highly relevant to mobile automation testing and require particular attention, but if handled correctly, then we can ensure a successful mobile testing effort.

Some of the factors that need to be taken care of in testing mobile applications are as follows:

- **Testing for cross device and platform coverage**: It is not feasible to test an application on each and every available device because of the plethora of devices that support the application across different platforms, which means you have to strategically choose only a limited, but sufficient set of physical devices. You need to remember that testing on one device, irrespective of whether it is of the same make, same operating system version, or uses the same platform cannot ensure that it would work on any other device. So, it is important that, at the very least, most of the critical features, if not all, are tested on a physical device. Otherwise, the application always runs a risk of potential failure on an untested device, especially when the target audience for the application is widespread, such as for a game or banking application.

  Use of emulated devices is one of the common ways to overcome the issues of testing on numerous physical devices. Although this approach is generally less expensive, we cannot rely completely on the emulated devices for the results they present, and with emulators, it may be quite possible that test conditions are not close enough to the real-life scenarios.

  So, an adequate coverage of different physical devices is required to test these following variations, providing sufficient coverage in order to negate the effects of fragmentation and have sufficient representation of these various factors:

  - Varying screen sizes
  - Different form factors
  - Different pixel densities and resolutions
  - Different input methods, such as QWERTY, touch screen, and more
  - Different user input methods, such as swipes, gestures, scrolling, and many more

- **Testing different versions of an operating system of the same platform**: For thorough testing, we need to test the application on all major platforms, such as Android, iOS, Windows, and others, for the target customer base, but each one of them has numerous versions available that keep on growing regularly. Most commonly, testing automation on the latest version of any operating system can be sufficient, as the operating systems are generally backward compatible. However, due to fragmentation of the Android OS, the application would still need to be tested on at least the most commonly used versions besides the latest ones, which in some cases may be significantly behind the latest version. This is because there may be many Android devices that are on an earlier version of Android and are not supported by the latest versions of Android.

- **Testing of various network types and network providers**: Most of the mobile applications, such as banking- or information-search-related applications require network connectivity, such as CDMA or GSM, at least partially, if not completely. If the application talks to a server about the flow of information to and fro, testing on various (at least all major) network providers is important. The network infrastructure used by network providers may affect data communication between application and the backend. Apart from the different network providers, an application needs to be tested on other modes of network communication, such as Wi-Fi network as well.

- **Testing for mobile-environment-specific constraints**: The mobile environment is very dynamic and has constraints, such as limited computing resources, available memory, in-between calls or messages, network switching, battery life, and a lot of other sensors and features, such as accelerometer, gyroscope, GPS, memory cards, camera, and others, present in the device, as an application's behavior depends on these factors. An application should integrate or interact (if required) with these features gracefully, and sufficient testing needs to be carried out in various situations to ensure this. However, oftentimes, it is not practically feasible to recreate all permutations and combinations of these factors, and hence a strategic approach needs to be taken to ensure sufficient coverage.

- **Testing for the unpredictability of a mobile user**: A tester has to be more cautious and should expand the horizon while testing the applications. They should make sure that an application provides an overall good response to all users and a good user experience; hence, **User Experience (UX)** testing invariably needs to be performed to a certain degree for all mobile applications. Since, a mobile application's audience comprises of various people ranging from nontech people to skilled technical users and from children to middle-aged users. Each of the users have their own style of using the application and have set their own expectations of it. A middle-aged or an aged user will be much calmer while using any application than someone who is young when it comes to the performance of the application. In general, we can say that mobile users have set incredibly high expectations of the applications available in the marketplace.

In the subsequent chapters, we will develop our understanding of how to overcome these challenges during automation design and how to strategize effectively. However, before that, we need to be aware of the different approaches for mobile test automation.

# Mobile automation testing approaches

In this section, you will understand the different approaches used for automation of a mobile application and their salient points.

There are, broadly speaking, four different approaches or techniques available for mobile application testing automation:

- Test automation using physically present real devices
- Test automation using emulators and simulators
- Mobile web application test automation through the user agent simulation technique
- Cloud-solutions-based test automation

# Automation using real devices

As the name suggests, this technique is based on the usage of real devices that are physically present with the testing automation team. Since this technique is based on the usage of real devices, it is a natural consequence that the **Application Under Test (AUT)** is also tested over a real network (GSM, CDMA, or Wi-Fi). To establish connectivity of the automation tool with the devices, any of the communication mechanisms, such as USB, Bluetooth, or Wi-Fi can be used; however, the most commonly used and the most reliable one is the USB connection. After the connection is established between the machines on which the automation tool is installed and the **Device Under Test (DUT)**, the automation scripts can capture object properties of the AUT and later, the developed scripts can be executed on other devices as well, but with minor modifications.

There are numerous automation tools, both licensed as well as open source freeware, available for mobile automation. Some commonly used licensed tools are:

- Experitest SeeTest
- TestPlant eggPlant Mobile /eggOn
- Jamo Solutions M-eux Test
- ZAP-fiX

Prominent tools for Android and iOS automation are:

- Selenium with solutions such as Selendroid and Appium along with iOS and Android drivers
- MonkeyTalk (formerly FoneMonkey)

The following are the salient features of this approach:

- The AUT is accessed on devices either by using a real mobile network or Wi-Fi network and can also be accessed by the Intranet network of the machine to which it is connected
- The automation testing tool is installed on the desktop that uses the USB or Wi-Fi connectivity to control devices under test

# Steps to set up automation

For automation on real devices, scripts are required to be executed on the devices with a USB or Wi-Fi connection to send commands via the execution tool to the devices. The following is a step-by-step description of how to perform the automation on real devices:

1. Determine the device connectivity solution (USB or Wi-Fi connectivity) based on the available setup. In some cases, USB connectivity is not enabled due to security policies and only in those cases is a Wi-Fi connection utilized.

2. Identify the tool to be used for the automation based on the tool feasibility study of the application. We will discuss in detail how to conduct a mobile automation tool feasibility study and the parameters that should be considered in subsequent chapters, when we discuss this technique in detail.

3. Procure the required licenses (seat or concurrent) if a licensed tool is selected. License procurement might mean that lengthy agreements need to be signed by both parties, besides arranging for the payment of services such as support. So, this step should be well planned with enough buffer time.

4. If the existing automation setup is to be leveraged, then an additional license needs to be acquired that corresponds to the tool (such as Quick Test Professional, Quality Center, and more). In some cases, you might also have to integrate the existing automation scripts developed with tools such as Quick Test Professional/Unified Functional Testing along with the automation scripts developed for the mobile. In such a case, the framework already in place needs to be modified.

5. Install the tools on the automation computer and establish the connectivity with the real devices. Installation may not be as simple as just running an executable file when it comes to mobile automation. There are various network-level settings and additional drivers that are needed to connect the computer and to control various mobile devices from the computer. Hence, all this should be done and planned well in advance.

6. Script the test cases and execute them on real devices.

# Limitations of this automation

This approach has the following limitations:

- The overall cost can be high as multiple devices are required to be procured for different teams and testers

- Maintenance and physical security can be an overhead

- Script maintenance can be delayed if testing cycles are overlapping with functional and automation teams

# Emulators-based automation

Emulators are programs that replicate the behavior of a mobile operating system and, to some extent, the device features on a computer. So, in essence, these programs are used to create virtual devices. So, any mobile application can be deployed on such virtual devices and then tested without the use of a real device. Ideally speaking, there are two types of mobile device virtualization programs: emulators and simulators.

From a purely theoretical standpoint, the following are the differences between an emulator and a simulator.

A device emulator is a desktop application that emulates both the mobile device hardware and its operating systems; thus, it allows us to test the applications to a lesser degree of tolerance and better accuracy. There are also operating system emulators that don't represent any real device hardware, but rather the operating system as a whole. These exist for Windows Mobile and Android, but a simulator is a simpler application that simulates some of the behavior of a device, does not emulate hardware, and does not work over the real operating system. These tools are simpler and less useful than emulators. A simulator may be created by the device manufacturer or by some other company that offers a simulation environment for developers. Thus, simulator programs have lesser accuracy than emulator programs.

 For the sake of keeping the discussion simple, we will refer to both as emulators in this chapter and in later chapters; when we discuss this technique in detail, we will refer to each of them individually.

Since this technique does not use real devices, it is a natural consequence that the AUT is not tested over a real network (GSM, CDMA, or Wi-Fi), and the network connection of the machine is utilized to make a connection with the application server (if it connects to a server, which around 90 percent of mobile applications do). Since the virtual devices are available on the computer, there is no external connection required between the device's operating system and automation tool. However, an emulator is not as simple as automating any other program because the actual AUT runs inside the *shell* of the virtual device. So, a special configuration needs to be enabled with the automation tools to enable the automation on the virtual device.

The following is a diagram depicting an Android emulator running on a Windows 7 computer:

In most projects, this technique is used for prelaunch testing of the application, but there are cases where emulators are automated to a great extent. However, since the emulator is essentially more limited in scope than the real devices, mobile-network-specific and certain other features such as memory utilization cannot be relied upon while testing automation with emulators. There are numerous automation tools, both licensed as well as of an open source freeware available for mobile automation on these virtual devices, and ideally, emulators for various mobile platforms can be automated with most of the tools that support real device automation.

The prominent licensed tools are:

- ExperiTest SeeTest
- TestPlant eggPlant Mobile / eggOn
- Jamo Solutions M-eux Test

Tools such as Selenium and ExperiTest SeeTest can be used to launch device platform emulators and execute scripts on the AUT.

The prominent free-to-use tools for emulator automation are:

- Selenium WebDriver
- Appium
- MonkeyTalk (formerly FoneMonkey)

Since emulators are also software that run on other machines, device-specific configurations need to be performed prior to test automation and have to be handled in the scripts. The following is the conceptual depiction of this technique.

The emulator and simulator programs are installed on a computer with a given operating system, such as Windows, Linux, or Mac, which then virtualizes the mobile operating system, such as Android, iOS, RIM, or Windows, and subsequently, which can be used to run scripts that emulate the behavior of an application on the real devices.

# Steps to set up automation

The following are the steps to set up the automation process for this approach:

1. Identify the various platforms for which the AUT needs to be automated.
2. Establish the connectivity to AUT by enabling the firewall access in the required network for mobile applications.
3. Identify the various devices, platforms, emulators, and device configurations, according to which test needs to be carried out.
4. Install emulators/simulators for the various platforms.
5. Create scripts and execute them across multiple emulators/simulators.

# Advantages

This approach has the following advantages:

- Standalone emulators that don't have real devices can be utilized
- No additional connectivity is required for automation
- This provides support for iOS and Android with freeware
- This provides support to all platforms and types of applications with licensed tools, such as Jamo Solutions M-eux and ExperiTest SeeTest

# Limitations

This approach has the following limitations:

- This can be difficult to automate as the emulators and simulators are themselves not thoroughly tested software and might have unknown bugs.

- Selenium WebDriver cannot be used to automate Android applications in some versions due to a bug in the Android emulator.

- It might sometimes be difficult to triangulate a defect that is detected on a virtual device and it might be needed that you recreate it on a real device first. In many cases, it has been observed that defects caught on emulators are not reproduced on real devices.

- For iOS simulators, access to a Mac machine with Xcode is required, which can be difficult to set up in a secure **Offshore Development Center (ODC)** due to security restrictions.

# User agent-simulation-based automation

The third technique is the simplest of all. However, it is also very limited in its scope of applicability. It can be used only for mobile web applications and only to a very limited extent. Hence, it is generally only used to automate the functional regression testing of mobile web applications and rarely used for GUI validations.

User agent is the string that web servers use to identify information, such as the operating system of the requester and the browser that is accessing it. This string is normally sent with the HTTP/HTTPS request to identify the requester details to the server. Based on this information, a server presents the required interface to the requesting browser.

This approach utilizes the **browser user agent manipulation** technique. This is depicted in the following schematic diagram:

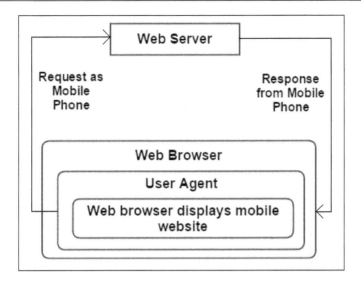

In this approach, an external program or a browser add-on is used to override the user agent information that is sent to the web application server to identify the requestor system as a mobile instead of its real information. So, for example, when a web application URL such as `https://www.yahoo.com` is accessed from a mobile device, the application server detects the requester to be a mobile device and redirects it to `https://mobile.yahoo.com/`, thereby presenting the mobile view. If the user agent information is overridden to indicate that it is coming from a Safari browser on an iPhone, then it will be presented with the mobile view.

The following screenshot depicts how the application server has responded to a request when it detects that the request is from an iPhone 4 and is presented the mobile view:

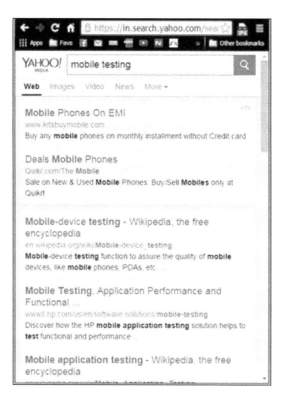

Since the mobile web application is accessed entirely from the computer, automation can be done using traditional web browser automation tools, such as **Quick Test Professional/Unified Functional Testing** or Selenium.

The salient features of this technique are as follows:

- With browser user agent manipulation, any mobile platform can be simulated
- Browser user agent manipulation is limited to only mobile web applications and is not extended to native and hybrid applications
- Browser simulation can be done using freeware files that are available for all leading web browsers

- The common user agent switching tools are:

    ○ Bayden UAPick for IE

    ○ User agent switcher add-on for Firefox

    ○ Fiddler for IE

    ○ Modify Headers for Firefox

    ○ UA Spoofer add-on for Chrome

    ○ Built-in device emulator with Chrome that can be accessed from developer tools

# Steps to set up the automation

The following are the steps to set up the automation process for this approach:

1. Identify the various platforms for which the AUT needs to be validated.

2. Identify the user-agent switcher tool that corresponds to any browser that needs to be leveraged for testing.

3. Identify the user-agent string for all platforms in scope and set up configuration in the user-agent switcher tool.

4. Leverage any functional testing tool that offers testing capabilities using any web browser, for example, Quick Test Professional, RFT, SilkTest, and Selenium WebDriver.

# Advantages

This approach has the following advantages:

- All platforms can be automated with little modification of scripts

- Quick implementation of automation solution

- Can leverage an open source software, such as Selenium for automation

- Existing automation set up can be leveraged

## Limitations

This approach has the following limitations:

- Least representative of real device-based tests
- Device-specific issues cannot be captured through this approach
- This cannot be used for UI-related test cases
- This approach supports only web-based mobile applications

# Cloud-based automation

This technique provides most of the capabilities for test automation, but is also one of the more expensive techniques. In this technique, automation is done on real devices connected to real networks that are accessed remotely through cloud-based solutions, such as **Perfecto Mobile**, **Mobile Labs**, **Sauce Labs**, and **DeviceAnywhere**.

The salient features of this technique are as follows:

- Cloud-based tools, such as Perfecto mobile and Device Anywhere provide a **WYSIWYG (What You See Is What You Get)** solution for automation
- Both **OCR (Optical Character Recognition)** and native object recognition and analysis is utilized in these tools for automation
- These tools also provide simple high-level keywords, such as **Launch Browser**, **Call Me**, and many more that can be used to design test cases
- The scripts that are thus created need to be re-recorded for every new type of device due to differences between the interface and GUI objects
- Mobile devices are accessed via a web interface or thick-client, by teams in various regions
- The devices are connected to real networks that use Wi-Fi or various mobile network operators (AT&T, Vodafone, and more)
- The AUT is accessed via the Internet or through a secure intranet connection
- This approach provides offer integration with common automation tools such as Quick Test Professional/UFT and Selenium

# Steps to set up the automation

The following are the steps to set up the automation process for this approach:

1. Identify the various platforms and devices for which the AUT needs to be automated.
2. Establish connectivity to AUT by enabling the firewall access for mobile web applications.
3. Open an account with the chosen cloud solution provider and negotiate to get the licenses for automation or set up a private cloud infrastructure within your company premises.
4. Install the cloud service provider client-side software setup along with the automation plugin for the relevant tool of choice (UFT or Selenium).
5. Book the devices as per testing needs (this usage normally has a cost associated with it).
6. Create scripts and execute across multiple devices.

# Advantages

This approach has the following advantages:

- This allows us to test automation on multiple devices of various manufactures (hardware)

  For example: Samsung, Apple, Sony, Nokia

- A script can be executed on multiple mobile devices from the same manufactures (models)

  For example: Galaxy SII, Galaxy SIII, iPhone 4S, iPhone 5, iPad2

- Scripts can be tested on different platforms (software)

  For example: Android 2.3 - 4.4, iOS 4-8, Symbian, Bada

# Limitations

This approach has the following limitations:

- Network latency may be experienced
- Cost can be high as fees depends on device usage
- Setting up a private mobile lab is costly, but may be necessary due to an organization's security policies, particularly in legally regulated industries, such as BFSI organizations

# Types of mobile application tests

Apart from the usual functional test, which ensures that the application is working as per the requirements, there are a few more types that need to be handled with an automation solution:

- **Interrupt testing**: A mobile application while functioning may face several interruptions that can affect the performance or functionality of an application. The different types of interruptions that can adversely affect the functionality of an application are:
    - ○ Incoming calls and SMS or MMS
    - ○ Receiving notifications, such as **Push Notifications**
    - ○ Sudden removal of battery
    - ○ Transfer of data through a data cable by inserting or removing the data cable
    - ○ Network/Data loss or recovery
    - ○ Turning a Media Player off or on

    Ideally, an application should be able to handle these interruptions, for example, whenever an interruption is there, an application can go into a suspended state and resuming afterwards. So, we should design automation scripts in such a way that they can not only test these interrupts, but they can reliably also reproduce them at the requisite step of the flow.

- **UI testing**: A user interface for a mobile application is designed to support various screen sizes and hence, the various components of a mobile application screen appear differently or in some cases, even behave differently as per the OS or device make. Hence, any automation script needs be able to work with varying components and also be able to verify the component's behavior. Use of automation ensures that the application is quickly tested and the fixes are regression tested across different applications. Since UI is where the end users interact with the application, use of a robust automation suite is the best way to ensure that the application is thoroughly tested so that it rolls out to the users in the most cost-effective manner. A properly tested application makes the end user experience more seamless and thereby, the application under test is more likely to get a better star rating and its key to commercial success.

- **Installation testing**: Installation testing ensures that the installation process goes smoothly without the user facing any difficulty. This type of a testing process includes not only installing an application but also updating and uninstalling an application. Use of automation to install and uninstall any application as per the defined process is one of the most cost-effective ways to do this type of testing.

- **Form factor testing**: Applications may behave differently (especially in terms of user interface) on smartphones and tablets. If the test application supports both smartphones and tablets, it should be tested on both form factors. This can be treated as an extension to the UI testing type.

# Selection of the best mobile testing approach

While selecting a suitable mobile testing approach, you need to look at the following important considerations:

- **Availability of automation tools**: The availability of relevant mobile automation tool plays a big role in the selection and implementation of the mobile automation approach.

- **Mode of connection of devices**: This is one of the primary, if not the most important, aspect that plays a pivotal role in the selection of a mobile automation approach.

  There are different ways in which devices can be connected to the automation tools such as:

  - Using a USB connection
  - Using a Wi-Fi connection
  - Using a Bluetooth connectivity (only for a very limited set of tools)
  - Using localized hotspots, that is, having one device as a hotspot and other devices riding its network for access
  - Cloud connection
  - Use of emulators and simulators

All these approaches need specific configurations on machines, and with the automation tools, which may sometimes be restricted, any automation solution should be able to work around the constraints in various setups. The key consideration is the *degree of tolerance* of the automation solution. The four different approaches that we discussed earlier in this chapter have each got a different level of accuracy. The least accurate is the user agent-based approach because it relies just on a web browser's rendering on a Windows machine rather than a real device. The most accurate approach, in terms of closeness to the real-world situation, is the use of real devices. However, this approach suffers from restrictions in terms of scalability of the solution, that is, supporting multiple devices simultaneously. Use of emulators and simulators is also prone to inaccuracies with respect to the real-device features, such as RAM, screen resolutions, pixel sizes, and many more. While working with cloud-based solutions, a remote connection is established with the devices, but there can be unwanted signal delays and screen refresh issues due to network bandwidth issues.

So, any approach that is selected for automation should factor in the degree of tolerance that is acceptable with any automation suite. For example, for a mobile application that makes heavy usage of graphics and advanced HTML 5 controls, such as embedded videos and music, automation should not be carried out with an emulator solution, as the degree of accuracy would suffer adversely and usually beyond the acceptable tolerance limit.

Consider another application that is a simple mobile web application with no complex controls and that doesn't rely on any mobile-device-specific controls, such as camera controls, or touch screen sensitive controls, such as pinch and zoom. Such an application can easily be automated with the user agent-based approach without any significant impact on the degree of accuracy.

If an application uses network bandwidth very heavily, then it is not recommended to use the cloud-based approach, as it will suffer from network issues more severely and would have unhandled exceptions in the automation suite. Conversely, the cloud-based approach is most suitable for organizations that have geographically and logically dispersed teams that can use remotely connected devices from a single web interface. This approach is also very suitable when there are restrictions on the usage of other device connection approaches, such as USB, Wi-Fi, or Bluetooth. Although this approach does need additional tools to enable cloud access, it is a worthwhile investment for organizations that have a high need for system and network security, such as banking and financial organizations.

# Troubleshooting and best practices

The following best practices should ideally be followed for any mobile automation project:

- The mode of connectivity between the AUT, DUT, and computer on which the automation tool is installed should be clearly established with all the considerations of any organization's security policies. In most cases, there is no way to workaround to the absence of USB connectivity, other than to use cloud-based automation solutions. So, before starting a project, the physical setup should be thoroughly vetted.

- The various operating systems and versions, mobile equipment manufacturers, and different form factors that need to be supported with the application, and consequently, the automation solution should be designed to support all of them. However, if you start automating before identifying all the supported devices, then there would invariably be a lot of rework required to make the scripts work with other devices. Hence, automation scripts should be made for all supported OSes and devices right from the design stage.

- A user agent-based automation can only be implemented for mobile web applications. It is a cost-effective and quick way to implement solutions since it involves automation of just a few basic features. However, this technique should not be relied upon for validating GUI components and should always be accompanied with a round of device testing.

- If any simulation or emulation technique (user agent or emulators/simulators) is used for automation, then it should strictly be used for functional regression testing on different device configurations. Ideally, projects utilizing these solutions should also have a GUI testing round with real devices, at least for the first release.

- If a geographically-distributed team is to utilize the automation solution, for example, an offshore-onsite team that needs to use the same devices, then the most cost-effective solution in the long run is the cloud-based automation. Even though the initial setup cost of the cloud solution generally is the highest of the four techniques, since different teams can multiplex and use devices from different locations and so the overall cost is offset by using fewer devices overall.

- During the use of emulators/simulators, the automation scripts should be designed to trigger the virtualization program with the required settings for memory, RAM, and the requisite version of the operating system, so that there is no manual intervention required to start the programs before you trigger the execution. Also, this way, scripts can be triggered remotely and in an unmonitored way.

- Irrespective of the technique utilized, a proper framework should be implemented with the automation solution. In the next chapter, we will discuss the various automation frameworks and their specific differences for mobile automation solutions.

# Summary

In this chapter, we learned what mobile test automation is, what are the different mobile packages that are available, and what factors should be considered during mobile automation testing. We then moved on to learn the different types of approaches and selection of the best approach according to any specific project requirements. So, it is evident that with the use of automation to test any mobile application, a good user experience can be ensured with a defect-free software, with which a good star rating can be expected for the AUT. In the next chapter, we will understand techniques used to design various mobile automation frameworks.

# 2
# Designing Mobile Automation Frameworks

A scalable and robust automation solution should be built with a specific automation framework that addresses the needs of the particular project. A test automation framework can be understood as a set of rules that governs the project delivery. It encompasses various aspects, such as the design methodology, data storage rules, and folder structure besides other things, such as coding standards to be followed, naming conventions of various files, other components of the automation-solution-like object repository, and recovery scenarios that are specific to any particular tool. In this chapter, you will learn about the various automation frameworks that are followed in typical automation projects, and then, we will extrapolate them for mobile-specific requirements.

In this chapter, we will cover the following topics:

- How to select the best suitable tool
- Types of frameworks with examples
- Troubleshooting and best practices

There are four fundamental types of automation frameworks that we will look at in detail in the subsequent sections with a focus on mobile automation:

- The functional decomposition or test script modularity framework
- The data-driven framework
- The keyword-driven framework
- The hybrid framework

We will first discuss what each of these frameworks are with respect to functional testing automation and then elaborate on the necessary modifications for each of these.

Each of these frameworks can be applied in combination with any of the mobile automation approaches, which are:

- Test automation using physically present real devices
- Test automation using emulators and simulators
- Mobile web application automation through the user agent simulation technique
- Cloud-based solutions automation

As we discussed in the previous chapter, each of the automation approaches can be supported with different tool sets, so that the concepts explained are easily applied to all the different tool sets. However, there are a few specific changes necessary for open source and licensed tools. So, to completely understand this, we will take the implementation examples of two market leaders in their respective category — **HP UFT** as a proprietary licensed tool and **Selenium** as an open source tool.

Defining any automation framework effectively for a particular project is a three-step process:

1. Select an automation approach suitable to the **Application Under Test (AUT)**.
2. Select an appropriate automation tool for the project-specific requirement.
3. Define an automation framework with the required customizations for mobile automation.

 You learned the basics of automation approach selection in the previous chapter. So, before we move on to the details of frameworks, it is imperative that you understand how to select a suitable automation tool along with the automation approach.

# Selecting an automation tool

The success of a mobile-automated testing depends on the selection of the right testing tools. Though there are numerous tools available in the market that are easily accessible, it always takes time to evaluate and find the suitable tool or tools for the specific project requirements in order to avoid changes and additional project costs later.

The following are a few scenarios that will allow us to make a suitable choice according to the project requirement as well as to set up the mobile test automation:

- **Support of data-driven inputs by the tool**: We need to manage different sets of data to enable test execution with a different set of data. The availability of appropriate drivers provides us with the ability to work with flat files, spreadsheets, and database storages. It also reduces the effort required to build support in order to read input data from various sources from scratch, which is often prohibitively costly to develop.

- **Type of mobile application packages supported**: Most of the tools available in the market do not support all the types of application packages, such as native, hybrid, or web applications, at the same time. So, it becomes necessary that we choose several tools based on the requirements, which should ideally be avoided in order to have a cohesive solution.

- **The scripting methods supported**: The easiest approach to start mobile test automation of a project is using the **playback** or **recording** approach, which is also the fastest way to test a mobile. Script parameterization is the only thing that is required in addition, to perform a mobile test. However, this approach is not as powerful as the coding approach, which is required to develop different types of frameworks. Often, coding of the scripts allows the use of programming language constructs, and thus, it is necessary that the tool allows the selection of a coded approach along with a pure record and playback option.

- **Supported mobile operating systems and its versions**: Based on the given requirement specifications of the project, we have to select such a tool that supports target mobile operating systems, such as iOS, Android, Windows, and any other tool required by your own project. As different versions of operating systems are available in market for operating systems, it is also very important to check the supported versions of operating systems and, quite often, it is also important to understand the speed with which newer operating system versions are covered with any given tool. In some cases, tool vendors take a very long time to develop support for newer versions of operating systems, which might be a hindrance if the application under test needs to be quickly validated and certified for newer OS versions.

- **Support for integration of other tools**: Choosing a right tool that gets integrated with the other available tools is very important, as we need the complete integration of mobile automation testing infrastructure components, such as **Integrated Development Environment (IDE)**, configuration management, continuous integration, revision control, test management, the test framework, execution management, report generation, defect tracking, and many other tools, so as to build a cohesive mobile test automation framework.

- **Ability to automate without access to the application source code**: Some tools require that the source code of the application be *instrumented* with the tool-supplied libraries in order to enable the tool to recognize the application controls. This is an important consideration while selecting any tool, as it might be possible that access to the application source code is not allowed to testers due to security concerns or just from a quality assurance perspective. Ideally, the application should be tested in the same form as it is going to exist in the production environment. So, either there should not be any need to access the application source code to enable the test automation, or the instrumentation process should work on the installable package rather than the complete source code.

- **Programming language supported**: Selecting and allocating the right resources is also very important, apart from selecting the right tool. Selecting tool languages, such as Java, Python, C#, Ruby, and many more, that match the client requirements and the availability of the resources that have expertise in the technology is a very important consideration. Also, advanced languages allow many ready-to-use libraries that are available, thus bringing down the overall development cost of the project.

- **Object recognition method supported**: It is always advisable to use unique object identification to simplify test scripts maintenance and reduce the impact of changes. When there is some change in the application, one of the common challenges faced by testers is the impact of the change on object recognition. If it is easy to manage the recognition properties in the collected object library, using the chosen tool, we just have to evaluate how object recognition is done. Many tools that rely on techniques such as **Optical Character Recognition (OCR)** are quite cumbersome to maintain, as there is often a need to maintain different objects of the same type for different device orientations and screen resolutions.

- **Reporting features**: Getting the execution logs in the required format is also very important. After the script's execution, it is not only enough to get the result's pass or fail status, but equally important to be able to capture other details, such as a screenshot of the failed scenario, details of the actual failed step, and many more. There are many tools available in the market that allow us to capture the detailed execution video logs that eliminate the need to manually recreate the flow to detect which exact step failed and thus saves the time of both the testing and development teams.

- **Cost of the tools**: There are several free and open source tools available in the market. One of them is Selenium, but we cannot always rely only on the use of open source tools. If we are planning to use them, then we need to be sure about how stable the tool evolution is and how quickly those tools are upgraded to support the latest changes in technologies.

  As for the proprietary solutions, such as **Unified Functional Testing (UFT)**, formerly known as **QuickTest Professional (QTP)**, the price of the tools is one of the key points that are taken into considerations in the mobile automation **return on investment (RoI)** calculation. We have to check the licensing model and the cost of any additional add-ons, support, and updates required for mobile automation.

- **Ease of use**: Ideally, the tool should be ideally easy to implement and learn, as for a complex tool that utilizes newer technologies and tool-specific proprietary languages, there might be a need to hire new workers or to train the existing workers, which considerably increases the overall cost of implementation of the project.

To track all of this information, we may utilize a **tool evaluation template**. The following is the format of the tool evaluation template:

| Parameters | | | Tool 1 | | Tool 2 | |
|---|---|---|---|---|---|---|
| S No. | Description | Weightage | Rating (out of 10) | Comments | Rating (out of 10) | Comments |
| | | | | | | |
| | | | | | | |
| Total | | | 0 | | 0 | |

Now, let's understand this template with a sample comparison between two tools. So, the following sample will make us understand how to use this template effectively and efficiently.

The **Weightage** column represents the importance any team would give to any parameter.

The following is a guideline for giving ratings to any parameter in the **Weightage** column:

- **Weightage 6** is for critical features
- **Weightage 5** is for the must-have features
- **Weightage 4** is for very important features
- **Weightage 3** is for important features
- **Weightage 2** is for the nice-to-have features
- **Weightage 1** is for the features that are not important
- **Weightage 0** is for the features that are not needed

Based on the level of importance and tool rating, the total is arrived at as a final weighted sum using the following formula:

$$Final\ Rating = \sum (Weightage\ as\ per\ Level\ of\ Importance \\ * Individual\ Parameter\ Rating)$$

The final rating can then be used to identify the best-suited tool.

 The following template is just for reference and not an actual comparison between any tools.

**The scenario**: An IT company got a new mobile automations testing project and the project manager was asked to select the required tool, out of various tools available in the market, that satisfies the project and client requirements. So, this is how the project manager uses the tool selection template to make an objective choice:

| Parameters | | Weightage | Selenium | | UFT | |
|---|---|---|---|---|---|---|
| S No. | Description | 0 to 6 | Rating (0 to 10) | Comments | Rating (0 to 10) | Comments |
| 1 | Testing types supported | 6 | 6 | Functional | 8 | Functional and UI |
| 2 | License cost | 6 | 10 | Open source | 7 | Licensing cost |

| Parameters | | Weightage | Selenium | | UFT | |
|---|---|---|---|---|---|---|
| 3 | Object recognition | 5 | 4 | No, can be programmatically designed | 7 | Yes |
| 4 | Tool scripting language support | 3 | 7 | Ruby/Python/ Java/C#/ JavaScript | 5 | VBScript |
| 5 | Supported browsers | 2 | 8 | IE, Firefox, Chrome, Safari, and Opera | 7 | IE, Firefox, and Chrome |
| 6 | Supported operating systems | 5 | 8 | iOS, Android, Windows, and Blackberry | 7 | iOS, Android, and Windows |
| 7 | Application package supported | 3 | 5 | Native and web | 8 | Native, hybrid, and web |
| 8 | Agile automation capability | 0 | 7 | Yes | 4 | No |
| 9 | Cloud approach support | 3 | 6 | Yes | 3 | No |
| 10 | Maintenance effort required | 4 | 7 | Medium | 5 | Low |
| 11 | Tool installation effort | 2 | 4 | High | 8 | Low |
| 12 | Public cloud support | 1 | 7 | Yes | 7 | Yes |
| 13 | Virtualization support such as VMware, Virtual PC, Citrix Env, and others | 2 | 8 | Yes | 8 | Yes |

| Parameters | | Weightage | Selenium | | UFT | |
|---|---|---|---|---|---|---|
| 14 | Testing approaches that can be used | 3 | 6 | Physical devices, emulators, and simulators | 9 | Physical devices, emulators and simulators, user-agent-based solutions and cloud-based solutions |
| Weighted total | | | 303 | | 308 | |

As we can see in the preceding table, using such a tool selection template, the project manager or architect can easily select UFT as the tool of choice for the required project, which is as per the evaluation of various parameters, on the basis of a final score achieved by the tool, 308, which is much higher than Selenium. As per the specific project requirements, the same parameter may be rated at different levels. Also, different teams may go for different tools even while comparing the same tools, as the weightage they may give to the parameters may be different and as per their own needs.

# Types of mobile automation frameworks

After the selection of an automation approach and appropriate tool as per the project requirements, we move on to the final step of selecting the appropriate framework to create a mobile automation solution for a project. In this section, we will discuss the various automation frameworks with a focus on mobile automation specifically.

# The functional decomposition or test script modularity framework

Test modularity or functional decomposition is an approach in which the application under test is divided into modules or script components for each of the unique functionalities of the AUT. This framework utilizes the basic concept of object-oriented programming to create abstract- or high-level components to hide the lower-level components. This insulates the component from modifications made in the application objects or flow changes and provides modularity in the automation suite design.

So, by applying the principles of abstraction and encapsulation, this framework improves the maintainability and scalability of automated test suites. Moreover, since we divide an AUT into modules, changes in any module do not affect other modules. With this approach, we can also easily reuse the functional components, and thus, the maintenance effort is reduced.

Using the functions and procedures, small and independent scripts that represent modules, sections, and functionality of the application under test are created. These small scripts are then used in a hierarchical fashion to construct larger tests, realizing a particular test case. Let's take a look at the following block diagram:

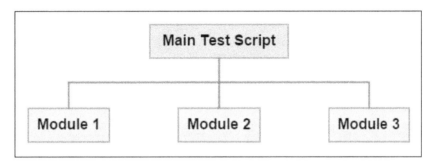

As an example, consider an e-mail application as your application under test. In the absence of a framework, the scripts would be linear blocks of code for various test cases. So, most probably, there would be a few initial lines of code at the beginning of every test case to start the browser and launch the e-mail application. This would introduce redundancy of code and with every increase in the number of test cases. This same block of code would be repeated in every script, resulting in a very high amount of maintenance if the launching method of the application is changed in any future release.

So, a basic method to overcome this redundancy is to follow the **object-oriented programming (OOP)** concept during the automation of scripts. Following this approach, you would need to develop various *modules* or *functions* that represent the repeatable steps that can later be reused in other test cases.

The following are some functions or modules that can be reused across different test cases:

 The sample code is provided for Selenium, which only explains the concepts. The same structures can be implemented for UFT as well with minor tool- and language-specific modifications.

```
Void fn_gotoEmailAccount() {
  'Open Browser and navigate to email application URL
  WebDriver driver = new InternetExplorerDriver();
  driver.get("https://testingemail.sit.com");
  'Page Sync
  driver.manage().timeouts().implicitWait(10,TimeUnit.SECONDS);
  'Click on create account
  driver.findElement(By.linkText("Create Account")).click();
}

'Enter the details
void fn_EnterDetails() {
  driver.findElement(By.name("First
  Name")).sendKeys("TestUserID1");
  driver.findElement(By.name("Last
  Name")).sendKeys("TestUserPwd123");
}

void fn_SubmitToCreate() {
  'Submit
  driver.findElement(By.name("Next Step")).click();
}
```

Now, the scripts would be created by calling these various steps as functions from the main script body, as follows:

```
'Test Script 1: To validate that application Login is allowed with
valid credentials
'Open EMail
fn_gotoTestingEMailAccount();
'Enter the details
fn_EnterDetails();
'Submit
fn_SubmitToCreate();
```

So, in this framework, we have two major components:

- **The function library**: This consists of the actual functions that represent the application functionality

- **The scripts library**: This calls the respective functions from the function library to create the actual test scripts

# Implementation required for mobile testing

The most important aspect that we need to consider while working with mobile automation is that not only should our framework be easily maintainable, it should also possess the ability to execute scripts across different devices (depending on the approach selected, this could be real devices, user agents, simulators, or cloud-connected devices). As we discussed previously, we also need to ensure that the scripts developed in any automation framework are able of getting executed on any given device and application type and should be scalable to add newer devices and application enhancements.

So, along with the function library, we also need to have a repository of the mobile devices (or their respective connection parameters, such as user agents) saved as global variables that can be manipulated at runtime and impact the entire set of scripts simultaneously. This way, all scripts can be executed on a single device at once without requiring repeated modifications for execution across different devices. Let's take a look at the following block diagram:

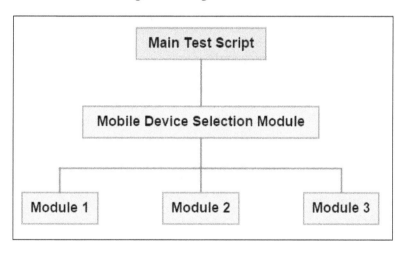

In the preceding diagram, a separate module should be maintained only for the mobile device information (launching, user agent, and so on) and should be connected at a global level to the framework box. Let's take a look at the following code snippet:

```
import junit.framework.TestCase;
import org.openqa.selenium.By;
import org.openqa.selenium.WebElement;
```

```
import org.openqa.selenium.android.AndroidDriver;
public class DriverLocator {
  public static WebDriver searchForPhone(String MobOSType) {
    WebDriver driver =null;
    switch(MobOSType) {
      case Android:
      driver = new AndroidDriver();
      break;
      case IPhone:
      driver = IPhoneDriver();
     break;
    }
  }
}
```

The `DriverLocator` class performs the function of switching and initiating the
required mobile driver type as per the `MobOSType` variable. In the case of Android,
the program instantiates an Android Driver and in the case of iPhone, it instantiates
an iPhoneDriver:

```
public class OneTest extends TestCase {
  public void testGoogle() throws Exception {
    WebDriver driver = DriverLocator.searchForPhone("Android");
    // And now use this to visit Google
    driver.get("http://www.google.com");
    // Find the text input element by its name
    WebElement element = driver.findElement(By.name("q"));
    // Enter something to search for element.sendKeys("Packt");
    // Now submit the form. WebDriver will find the form for us
    from the element
    element.submit();
    // Check the title of the page
    System.out.println("Page title is: " + driver.getTitle());
    driver.quit();
  }
}
```

The `OneTest` class then creates a test case by calling the `searchForPhone()` function
with the `Android` parameter and using the Android Driver to launch a simple flow
with the `www.google.com` website to search for the `Packt` parameter and retrieve the
page title.

It is evident that for another test case that needs the same flow to be tested over
an iPhone, a new test case needs to be developed that calls the `searchForPhone()`
function with the `IPhone` parameter.

So, with the test script modularity or functional decomposition framework, although we can take advantage of reusability of components and reduced maintenance, it suffers from the limitation of lack of data reusability as the data is hardcoded within each component. Hence, we can't use the same component with multiple data values without changing the values before each execution or having multiple test cases with the same flow but different data. This limitation can be overcome with the use of the data-driven framework, which we will discuss now.

# The data-driven framework

This is a framework where the test input and output values are loaded into variables in captured or manually coded scripts and are read from data files, such as CSV, DAO objects, ODBC sources, Excel files, ADO objects, and others. Information such as reading of the data files, logging of the test status, and navigating through the programs are coded in the test scripts. In this type of framework, variables are used for both input values and output verification values.

So, the information of the test data to be used in the test case is contained in a data file and is not directly a part of the script. The script can be considered as a *driver* for the data to be fed in the repeatable steps.

This framework, when it comes to developing workarounds for bugs and performing maintenance, offers the greatest flexibility. This framework also tends to reduce the overall number of scripts we need in order to implement all of our test cases.

For example, if we've a need to validate the login functionality of the `testingEmail` application mentioned in the previous section, we would need to use various combinations of the test data for the `User ID` and `Password` values. It can be correct ID/correct password, correct ID/incorrect password, correct ID/blank password, blank ID/correct password, and so on. As we can understand, the same steps of entering a User ID, Password, and pressing the **submit** button press are repeated for all these test cases. If we create a modular framework, separate functions for the correct ID/incorrect password and the various combinations mentioned previously would be required. Hence, the overall number of functions or modules would be increased. A simple solution is to keep the test data for the **User ID** and **Password** fields in a data source, such as an Excel sheet, and then repeat the same steps with this different data. This method of keeping data in various source files and using them during execution of the same steps repeatedly is referred to as **Parameterization**.

Many tools such as UFT provide an inbuilt capability to link input files and read the data using predefined functions.

In the following example, the code for a UFT script, we see that the implementation of the test data with the hardcoded values in a test script modularity framework is used to log in to the Gmail application:

```
Browser("Gmail: Email from
Google").Page("GoogleAccounts").WebEdit("FirstName").Set
"tester_Name"
Browser("Gmail: Email from Google").Page("Google
Accounts").WebEdit("LastName").Set "Test_LastName"
Browser("Gmail: Email from
Google").Page("GoogleAccounts").WebEdit("GmailAddress").Set
"Tester1@gmail.com"
```

The hardcoded data can be easily moved to a data input sheet using the parameterized variables:

```
Browser("Gmail: Email from Google").Page("Google
Accounts").WebEdit("FirstName").Set DataTable("First_Name",
dtGlobalSheet)
Browser("Gmail: Email from Google").Page("Google
Accounts").WebEdit("LastName").Set DataTable("Last_Name",
dtGlobalSheet)
Browser("Gmail: Email from Google").Page("Google
Accounts").WebEdit("GmailAddress").Set DataTable("Email_Address",
dtGlobalSheet)
```

With an Excel file that has three inputs for the three different variables, the script will repeatedly get executed three times with different input combinations instead of creating three different scripts, such as shown in the following table:

| First_Name | Last_Name | Email_Address |
| --- | --- | --- |
| Tester1 | Worker_1 | Tester1@email.com |
| Tester2 | Worker_2 | Tester2@email.com |
| Tester3 | Worker_3 | Tester3@email.com |

So, a data-driven framework typically has the following components:

- **Test scripts**: These are the repetitive flows of steps to be iterated on the required test data
- **Data files**: These are the data files, such as CSV, Excel, MS Access, and so on, that contain the data

Let's take a look at the following block diagram:

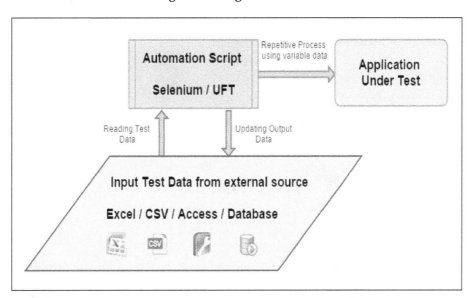

The data-driven framework stores the information of the application's test data in external files

# Implementation required for mobile testing

In this type of framework implementation for mobile testing, we can treat the information for various mobile devices (user agents, device connection parameters, and emulator instances) as test data. However, it should be treated separately from the application test data. Generally, in a data-driven framework, the input data is linked to a specific step in the script and that is where the problem lies. If the information of the device is linked to a specific step in the test case, then it needs to be individually modified every time the test suite is required to be executed on different devices. A solution to this problem is that for mobile implementation, the data for the required mobile devices should be linked to the *Launch Device* step only, but should be referenced as a variable, so that the device-specific changes to the application can be handled without referring to the mobile device data directly. This way, hardcoding of the device information is avoided.

Another more elegant approach is to keep a separate input data file for the device-specific information and to reference the device information in the input data in order to launch the device component as a variable name, such as shown in the following block diagram:

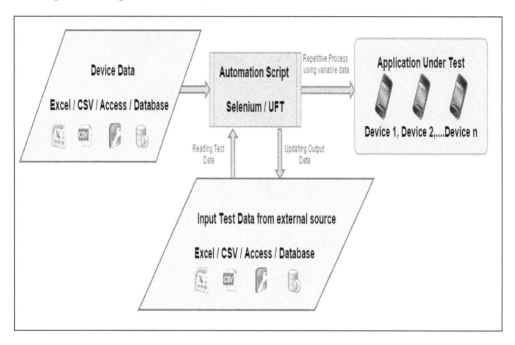

As seen in the preceding diagram, the mobile data-driven framework has a separate input sheet for the device-specific information.

Let's take a look at the following code snippet, which gets information about the different mobile devices, on which the test needs to be performed, from the Excel sheet:

```
public class DataTable {
  static Connection c1;
  public static void  getConnection() {
    try {
      Class.forName( "sun.jdbc.odbc.JdbcOdbcDriver");
      Coonection c1= java.sql.DriverManager.getConnection(
      "jdbc:odbc:Driver={Microsoft Excel Driver
      (*.xls)};DBQ="+pathToDataTableFile+";READONLY=FALSE");
    }
    catch(Exception e) {
      e.printStackTrace();
    }
```

```
  }// End of getConnection
  public static String getDeviceData(String strColumnName) {
    String strColumnVal =null;
    try {
      String query = ""SELECT DeviceType FROM ["DeviceData"$]
      where TC_ID ='"+testcase+"'";
      Statement st = c1.createStatement();
      ResultSet rs = st.executeQuery(query);
      strColumnVal = rs.getString(strColumnName);
    }
    catch(Exception e) {
      e.printStackTrace();
    }
    return strColumnVal
  }// End of getColumnData
}
```

The DataTable class is required to connect and read the data table that is contained in an Excel file named DeviceData, as shown in the following table, which contains the DeviceData Excel sheet containing the device specific information for each test case:

| Test Case ID | Device Type |
|---|---|
| TC1 | Android |
| TC2 | iPhone |
| TC3 | Android |

Let's take a look at the following code snippet, which finds out from which device the user has logged in on Gmail:

```
import org.openqa.selenium.By;
import org.openqa.selenium.WebDriver;
import org.openqa.selenium.android.AndroidDriver;
import org.openqa.selenium.iphone.IPhoneDriver;
public class Gmail {
  public static void main(String args[])throws Exception {
    WebDriver ad = null;
    Switch (DataTable.getDeviceData("DeviceType")){
      case Android:
      ad = new AndroidDriver();
      break;
      case IPhone:
      ad = new IPhoneDriver();
```

```
        break;
    }
    System.out.println("Started");
    ad.get("http://www.gmail.com");
    System.out.println("Application Title"+ ad.getTitle());
    Thread.sleep(2000);
/* The switch statement selects the corresponding driver as per
DeviceType column in the data table for the application specific
data*/
    ad.findElement(By.name("FirstName")).
    sendKeys(DataTable.getColumnData("First_Name"));
    ad.findElement(By.name("LastName")).
    sendKeys(DataTable.getColumnData("Last_Name"));
    ad.findElement(By.name("GmailAddress")).
    sendKeys(DataTable.getColumnData("Email_Address"));
    ad.findElement(By.name("signIn")).click();
    org.testng.Assert.assertEquals(ad.getTitle(), "Gmail
    Account");
//The assert statement would report as success or failure in the
result log
    System.out.println("Opened");
    ad.quit();
    }
}
```

So, with the data-driven framework, the main advantage is we can repeat the execution of the similar steps on the variable data. However, for complex applications where there are multiple application functionalities and there is a need to reuse the same set of data for different test cases, a purely data-driven framework cannot be utilized.

# The keyword-driven framework

The keyword-driven framework is also sometimes referred to as the table-driven testing framework. In this framework, the application functionality is broken down into *modules* or *functions* just as in the test script modularity framework, which are referred to as keywords. However, the difference between the keyword-driven and the modular framework is that while in the modular framework the scripts are created by calling one function after another from a separate library file, in the keyword-driven framework, the scripts are created by referring them from an input table.

The test automation tool is used to execute the scripts, and the test script code that drives the application under test and the various keywords are referred to as the *Driver script*. So, when you look at the *scripts* in a keyword-driven framework, they are normally very similar to how manual test cases are written with high-level information in a tabular format in an external file, such as an Excel sheet, CSV, or a database file.

To get an insight into this framework, let's consider a scenario where we have to develop an automation suite for a banking application that provides various services, such as account balance check, making transactions, paying bills, and scheduling payments for utility bills. Here, the basic functionalities of the application can be understood as keywords representing the application features. So, the automation framework would have keywords such as **Login**, **Account_Balance_Check**, **Fund_Transfer**, **Utility_Bill_Pay**, and **Logout**. All the test cases in the automation suite would be created using various combinations of these basic functionalities only. Hence, if we are able to develop a keyword library of the application functionalities, then the entire suite can be created by using them in proper combinations.

While using the modular framework, each automation script is part of another file, and hence, a lot of effort is needed to maintain the test cases along with the expertise of automation tools. Whereas, with the keyword-driven framework, since we utilize tables that are easy to understand, the scripts can be kept external to the automation library and even functional testers can then understand the flow and make their own test cases using these keywords.

The keyword implementation of such a test suite would look like the one shown in the following table:

| Script ID | Keyword 1 | Keyword 2 | Keyword 3 | Keyword 4 |
|-----------|-----------|-----------|-----------|-----------|
| TC_1 | Login | Account_Balance_Check | Fund_Transfer | Logout |
| TC_2 | Login | Fund_Transfer | Account_Balance_Check | Logout |
| TC_3 | Login | Account_Balance_Check | Utility_Bill_Pay | Logout |
| TC_4 | Login | Utility_Bill_Pay | Logout | |

So, we have now created four test cases by combining the various *keywords* that represent the *functionalities* of the application in different sequences for each test case. Since the application's entire functionality is covered in these keywords, all test cases can now be successfully created and executed, ensuring complete coverage of the test cases with minimal effort.

Based on the tool with which this framework is implemented and the type of files used to keep the test case data, different driver files are developed, but the basic structure remains the same. Let's take a look at the following block diagram explaining the use of the **Keyword Library**:

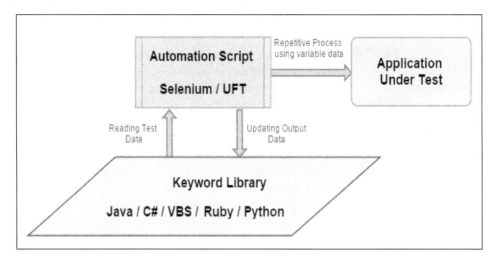

## Implementation required for mobile testing

In this type of framework implementation for mobile test, we can treat the information for various mobile devices as a specific keyword that handles the required device-specific information for the selected approach (user agent, device emulator, or cloud connection parameters). After this, the execution can be controlled over multiple devices from a single keyword or parameter of the framework. By calling different mobile-device-specific keywords, the rest of the keywords can be executed on one device. Let's take a look at the following mobile devices and the keywords that are associated with them:

| Script ID | Keyword 1 | Keyword 2 | Keyword 3 |
|-----------|-----------|-----------|-----------|
| TC_1 | LaunchApp_Android_ SamsungS2 | Login_Android | Logout_Android |
| TC_2 | LaunchApp_iOS_iPhone5 | Login_iPhone | Logout_iPhone |
| TC_3 | LaunchApp_iOS_iPadMini | Login_iPad | Logout_iPad |
| TC_4 | LaunchApp_Windows_Lumia | Login_ Windows | Logout_Windows |

Here, in the script with the **TC_1** ID, the **LaunchApp_Android_SamsungS2** keyword is used to start an Android Samsung S2 device and the remaining keywords, **Login** and **Logout**, then get executed over the launched device. The following diagram shows the entire process that takes place during automation process:

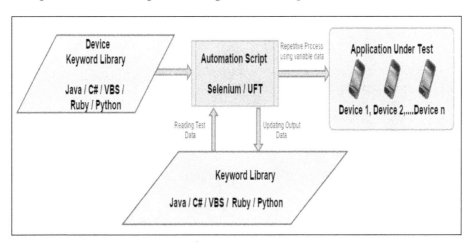

Let's take a look at the following code snippet:

```
public class BusinessFlowReader {
    static Connection c1;
    public static void  getConnection() {
        try {
            Class.forName( "sun.jdbc.odbc.JdbcOdbcDriver");
            Connection c1= java.sql.DriverManager.getConnection(
            "jdbc:odbc:Driver={Microsoft Excel Driver
            (*.xls)};DBQ="+pathToBusinessFlowFile+";READONLY=FALSE");
        }
        catch(Exception e) {
            e.printStackTrace();
        }
    }
    public String[] getBusinessComponentsKeyword(string testCase) {
        String[] arrKeywords=null;
        try {
            String query = ""SELECT * FROM ["BusinessFlow"$] where TC_ID
            ='"+testCase+"'";
            Statement st = c1.createStatement();
            ResultSet rs = st.executeQuery(query);
            rs.next();
            int colCount=rs2.getMetaData().getColumnCount();
```

```
       arrKewords = new String[colCount];
       int p,q;
       for(p=2,q=0;p<colCount;p++,q++) {
         if(rs.getString(p)==null)
         break;
         arrKewords[q]= rs.getString(p);
       }
     }
     catch(Exception e) {
       e.printStackTrace();
     }
     return arrKeywords;
     }

   }
 }
```

The Reflection class in the following code is used for the execution of the respective keywords as per the business flow mentioned in the BusinessFlow Excel file. The keywords array contains the list of all the keywords for a particular test case denoted by a row in the BusinessFlow Excel file. This class in turn initializes the BusinessComponents class and executes the methods for the required business flow:

```
public class Reflection {
  public static void execute(String keywords[]) throws Exception {
    Class c1 = Class.forName("BusinessComponents");
    Object o = c1.newInstance();
    for(int i=1;i<keywords.length;i++) {
      Method m = c1.getDeclaredMethod(keywords[i]);
      m.invoke(o,null);
    }
  }
}
```

The DeviceKeywordLibrary class file contains the functions for each of the device types for which execution needs to be done.

```
public class DeviceKeywordLibrary {
  Webdriver driver=null;
  void LaunchApp_Android_SamsungS2() {
    driver=new AndroidDriver();
    DesiredCapabilities cap = new DesiredCapabilities();
    cap.setCapability("version", deviceOSVersion);
    cap.setCapability("platform", deviceOS);
```

```
    WebDriver driver = new RemoteWebDriver(new
    URL("http://localhost:8089/wd/hub"),cap);
    driver.get("http://www.bankingTestingsite.com");
    Reporter.log("Application opened successfully");
  }

  void LaunchApp_iOS_iPhone5() {
    driver=new IPhoneDriver();
    driver.get("http://www.bankingTestingsite.com");
    Reporter.log("Application opened successfully");
  }
}
```

The `BusinessComponents` class file contains all the functions for each of the business components. The `Reflection` class is utilized to execute the corresponding business functions from this class:

```
public class BusinessComponents {
  void login() {
    String strUsername = "User01"
    String strPassword = "Password01"
    driver.findElement(By.name("UserId")).sendKeys(strUsername);
    driver.findElement(By.name("Password")).sendKeys(strPassword);
    driver.findElement(By.name("Submit")).click();
  }
  void accountBalanceCheck() {
    String accountBalance
    =driver.findElement(By.name("accountSummary")).getText();
    if(accountBalance.equals("10000")) {
      Reporter.log("Account balance is equal to 10000");
    }
    else {
      Reporter.log("Accoutn Balance is not equal to 10000");
    }
  }
  void logout() {
    driver.findElement(By.name("logout")).click();
    Reporter.log("Clicked on Logout link");
    Org.testng.Assert.assertEqual(driver.getTitle(),"Logged out");
  }
}
```

The `DriverScript` class triggers the execution flow as per the keywords from the corresponding test case row in the `Test Case` file:

```
pubic class DriverScript {
  public static void execute() {
    String keywords[] = DataTable. getBusinessComponentsKeyword();
    Reflection.execute(keywords);
  }
}
```

So, we can observe that this framework provides the advantage of easy configuration outside the tool and inbuilt modularization, making the development easier and the overall suite more maintainable. However, a purely keyword-driven framework does not handle the variable data, and thus, the test data is often required to be hardcoded in specific keywords.

Consider a scenario in the previous example, where different test data needs to be used for customers with savings and checking accounts. In that case, two separate keywords for the login functionality need to be developed—**Login_Savings_Account** and **Login_Checking_Account**. Although the same basic steps are to be repeated to log in, because of the difference in the test data, we would need to develop and maintain two different keywords. This increases the maintenance effort for a purely keyword-driven framework, and to address this shortcoming, we have an amalgamation of the advantages of all framework types called the hybrid framework, which we will discuss in the next section.

# The hybrid framework

A hybrid framework, as the name suggests, is a combination of various frameworks. It is a natural evolution of the various frameworks and blends the strengths of keyword and data-driven frameworks and eliminates their limitations. So, it is observed quite often that projects that run long, end up utilizing hybrid frameworks. By bringing together the principles of data-driven and keyword-driven frameworks, we can both eliminate redundancy and improve the reusability of test cases both at the same time. Hence, it is the most commonly implemented framework across all automation projects, and it is what most frameworks gradually evolve into over a period of time.

To understand the hybrid framework, let's continue with the example given while understanding the keyword-driven framework. It can be understood that if we combine the concept of *parameterization* from the data-driven framework, the savings or checking account type is considered as the input data for the login action, and hence, it can be moved into a data table for the **Login** keyword. This will result in only one keyword for the login functionality, which will take the account type information as the test data parameter and perform the necessary steps.

The following is a representation of the hybrid framework implementation for the previously mentioned example of the banking website:

| Script Id | Keyword 1 | Keyword 2 | Keyword 3 | Keyword 4 |
|---|---|---|---|---|
| TC_1 | Login_ Parameterized | Account_Balance_ Check | Fund_Transfer | Logout |
| TC_2 | Login_ Parameterized | Fund_Transfer | Account_Balance_ Check | Logout |
| TC_3 | Login_ Parameterized | Account_Balance_ Check | Utility_Bill_Pay | Logout |
| TC_4 | Login_ Parameterized | Utility_Bill_Pay | Logout | |

Refer to the Login() function from the previously explained example in the keyword-driven framework and note that, by parameterizing the input data, a more efficient implementation can be achieved that allows the use of the same **Login_ Parameterized** component for various input data as per the test case. The test data input sheet would have the following structure to store the **Account Type**, **Login ID**, and **Password** details:

| Script Id | Account Type | Login ID | Password |
|---|---|---|---|
| TC_1 | Savings | Savings_user1 | Tester1 |
| TC_2 | Checking | Checking_user1 | Tester2 |
| TC_3 | Checking | Checking_user2 | Tester4 |
| TC_4 | Savings | Saving_user2 | Tester3 |

So, the first script would utilize the information from the test data input sheet to perform the login action using the Login ID as **Savings_user1**, and thus, the test case would get executed for a savings account user. Thus, without increasing the number of **Login** keywords, the same action can be performed with varying test data. This improves the overall maintainability as well as the readability of the test suite, and even manual testers can then be involved to create their own test cases with different input data and combinations of keywords. Let's take a look at the following block diagram explaining the process of automation:

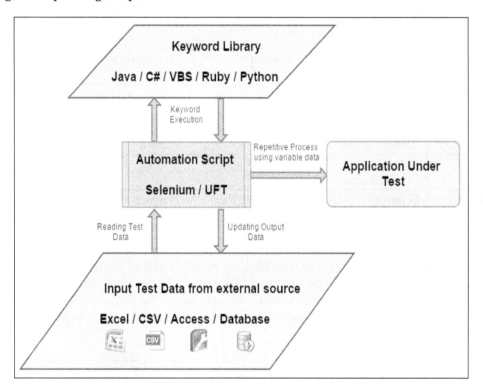

Apart from the same components as the ones present in the keyword-driven framework example, only the `Login` function in the `BusinessComponents` class needs to be parameterized. The following is the code for the **Login_Parameterized** component:

```
void login() {
    String strUsername = DataTable.getColumnData("Login ID");
    String strPassword = DataTable.getColumnData("Password");
    driver.findElement(By.name("UserId")).sendKeys(strUsername);
    driver.findElement(By.name("Password")).sendKeys(strPassword);
    driver.findElement(By.name("Submit")).click();
}
```

# Implementation required for mobile testing

As we did in the implementation of the keyword-driven framework, we can parameterize the mobile-specific information into a test data input table. However, care must be taken to keep the data input sheet for mobile-specific information separate from the test case input data sheet. Also, the mobile-specific information should not be linked to any specific test script ID. If we keep the mobile-specific information as part of the test data input sheet, then the mobile-specific information is repeated multiple times, and in case of any change to the input data, it would then need to be updated in multiple rows. This would lead to very high maintenance and would increase the chances of human error while updating the datasheets. The following table shows the deive ID's of different devices and the corresponding keywords that are assigned to it:

| Script Id | Device_ID | Keyword 1 | Keyword 2 | Keyword 3 |
|-----------|-----------|-----------|-----------|-----------|
| TC_1 | D1 | LaunchApp | Login | Logout |
| TC_2 | D2 | LaunchApp | Login | Logout |
| TC_3 | D3 | LaunchApp | Login | Logout |
| TC_4 | D4 | LaunchApp | Login | Logout |

The mobile-device-specific input sheet would be as follows:

| Mobile device ID | Device model | OS |
|------------------|--------------|----|
| D1 | Samsung S2 | Android |
| D2 | iPhone 5 | iOS |
| D3 | iPad mini | iOS |
| D4 | Lumia | Windows |

The device data Excel file containing the individual device details

It is important to note that the test data sheet would remain the same as the earlier one and would be linked to specific test script IDs, but there would now be an additional mobile input data sheet that would have data independent of test script IDs.

This mobile-specific data would need to be passed as a global parameter at the driver script level where it would be used as a parameter to the **LaunchApp** keyword, and all the scripts then can be executed on the same device, as shown in the following block diagram:

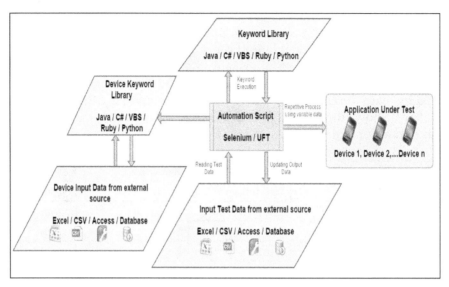

The hybrid framework implementation for mobile testing

If it is required that different test scripts should be executed on specific devices during an execution cycle, then **Device ID** can be used as a parameter to the **LaunchApp** keyword, as shown in following table:

| Script Id | Device ID | Keyword 1 | Keyword 2 | Keyword 3 | Keyword 4 |
|-----------|-----------|-----------|-----------|-----------|-----------|
| TC_1 | D1 | LaunchApp | Login_ Parametrized | Account_ Balance_ Check | Logout |
| TC_2 | D3 | LaunchApp | Login_ Parametrized | Fund_Transfer | Logout |
| TC_3 | D4 | LaunchApp | Login_ Parametrized | Account_ Balance_ Check | Logout |
| TC_4 | D2 | LaunchApp | Login_ Parametrized | Fund_Transfer | Logout |
| TC_5 | D3 | LaunchApp | Login_ Parametrized | Account_ Balance_ Check | Logout |

The BusinessFlow Excel file containing the individual test case details along with the device variable and parameterized keywords

So, in this suite, the first script would be executed on the Android Samsung S2 device and the second script would be executed on the Apple iPad mini. If in future the device parameters are changed or it is required that the test scripts developed for iPad mini be executed on the Samsung Galaxy tab instead, then this can be achieved simply by changing the data for the device ID **D3** in the device sheet, as follows:

| Mobile device ID | Device model | OS |
| --- | --- | --- |
| D1 | Samsung S2 | Android |
| D2 | iPhone 5 | iOS |
| D3 | Samsung Galaxy tab | Android |
| D4 | Lumia | Windows |

The device data Excel file can be easily modified externally to the automation scripts (change in D3)

Now, the test script IDs **TC_2** and **TC_5** would be executed on the Samsung Galaxy tab instead of the iPad mini with no other updates required on the actual code of test scripts or keywords.

The test data for the scripts would continue to be referred from the input data sheet, as shown in the following table:

| Script ID | Account type | Login ID | Password |
| --- | --- | --- | --- |
| TC_1 | Savings | Savings_user1 | Tester1 |
| TC_2 | Checking | Checking_user1 | Tester2 |
| TC_3 | Checking | Checking_user2 | Tester4 |
| TC_4 | Savings | Saving_user2 | Tester3 |

The test data input Excel file containing the test data for individual test cases

The `launchApp()` function in the following code snippet reads the values of various devices from the device information sheet and then uses that information to instantiate the required mobile device type on a cloud-based solution called Mobile Labs:

```
public class SupportComponents {
  static WebDriver launchApp() {
    String url = Datatable.getColummValue("dcUrl");
    String username = Datatable.getColummValue("dcUserName");
    String password = Datatable.getColummValue("dcPassword");
    String deviceId = Datatable.getColummValue("dcDeviceId");
    String scale = Datatable.getColummValue("dcScale");
    String orientation =
    Datatable.getColummValue("dcOrientation");
```

```
    String appID = Datatable.getColummValue("dcAppId");
    String command = "MobileLabs.deviceViewer.exe
    "+url+""+username+""+password+""+deviceId+""+scale+"
    "+orientation+""+appID;
    Runtime r = Runtime.getRuntime();
    Process p = r.exec(command);
    DesiredCapabilities cap = DesiredCapabilities.AndroidDriver();
    cap.setCapability("version","4.4");
    cap.setCapability("platform","Anndroid");
    WebDriver ad = new RemoteWebDriver(new
    URL("http://localhost:8089/wd/hub"),cap);
    ad.get("http://www.bankingsite.com");
    Report.LogInfo("App launch","application launched
    successfully",  Status.Done);
    Return ad;
  }
}

public class BusinessComponents {
  void login() {
    WebDriver ad = SupportComponents.launchApp();
    String strUsername = Datatable.getColumnValue("Userid")
    String strPassword = Datatable.getColumnValue("Password")
    ad.findElement(By.name("UserId")).sendKeys(strUsername);
    ad.findElement(By.name("Password")).sendKeys(strPassword);
    ad.findElement(By.name("Submit")).click();
  }
  void accountBalanceCheck() {
    String accountBalance =
    sd.findElement(By.name("accountSummary")).getText();
    if(accountBalance.equals("10000")) {
      Report.LogInfo("Balance Check","Account Balance is 10000",
      Status.Pass);
    }
    Else {
      Report.LogInfo("Balance Check","Account Balance is not
      10000",Status.Fail);
    }
  }
  void logout() {
    ad.findElement(By.name("logout")).click();
    Report.LogInfo("Logout","Logged out
    successfully",Status.Done);
  }
}
```

# Support libraries

The **support libraries** are the general-purpose routines and utilities that let the overall automation framework do what it needs to do. They are the modules that provide services such as:

- File handling
- String handling
- Buffer handling
- Variable handling
- Database access
- Logging utilities
- System\Environment handling
- Application mapping functions
- System messaging or system API enhancements and wrappers

The support libraries also provide traditional automation tool scripts access to the features of our automation framework, including the application map functions and the keyword-driven engine itself. Both of these items can vastly improve the reliability and robustness of these scripts until such time that they can be converted over to keyword-driven test tables.

There are many support frameworks for various automation tools and technologies available, such as **TestNG**, **Cucumber**, **JUnit**, **NUnit**, and others. These support frameworks can all individually be used to implement the four conceptual frameworks that we have discussed. These support frameworks provide many features, such as parallel execution and annotations for inbuilt reporting, that help to reduce the overall effort spent in implementation.

# Optimizing the combination of approach and framework

Just as we discussed, different mobile automation approaches can be utilized for a project. So, we can have a combination of automation frameworks implemented in the same project as per the requirements. It is not necessary that we always use the most sophisticated mechanism to automate, and quite often, we can implement an automation project with a combination of approaches as well as frameworks. For example, with the cloud-based automation approach, you may want to use a hybrid framework, but for the same application's web version, you may want to use a purely data-driven automation framework.

# A robust reporting mechanism

You must not always rely on the reporting mechanism purely provided by the automation tools and should build a reporting structure best suited to the automation framework. There are various ways to generate readable reports with UFT and Selenium, which should be utilized.

# The automation feasibility analysis

We have not touched upon this subject in great detail because we believe that you will be well versed with how to conduct a feasibility study for mobile automation on your respective projects. So, nowhere do we suggest that, despite the automation feasibility analysis, all mobile automation projects should always be automated. The final decision should be made based on the RoI aspect and technical feasibility of automating and maintaining the automation suite of a mobile application.

# Test library management

Test library files are required to be managed just as any other piece of software, but since the focus in automation projects is more on the development of scripts, this aspect is often overlooked and not included in the overall framework implementation/architecture. Ideally, there must be a common repository of all the scripts developed along with the data and any other relevant information utilized in the framework. All these files should be included and managed with change and version-control mechanisms, such as CSV, GIT, or SVN, so that changes are allowed only for the authorized users as well as concurrent updates are not performed.

# Version control

Proper version control of any test library allows you to perform test execution for the corresponding version of the application without any conflicting changes made to the test suite for application modifications in newer versions. This requires that for each application version, a corresponding version of the test library is maintained at the same time.

# Change control

While updating the application functionality or objects, the corresponding automation suite is also required to be maintained. Since this process is iterative, it often becomes cumbersome to maintain different versions of the automation suite for different environments of the application under tests without a proper change control mechanism to track them. Along with the tool for version control, a mechanism to track changes should also be developed in the form of change logs. If the change to a test is for a new feature or modification in a different application version, then the new test should be checked in a different version of the test library instead of overwriting the test for the older versions.

# Summary

In this chapter, you learned how to select the most suitable automation tool for a project, with the help of the tool selection template, as per the project-specific requirements. You learnt about the four fundamental types of automation frameworks with a focus on mobile automation. With the various diagrams, code snippets, and examples, you learnt about the frameworks in detail. You then learnt about the pros and cons of the various frameworks. You also learnt what are support libraries, and finally you learnt about the various best practices and troubleshooting exercises to enhance the capabilities. As you have learnt that, eventually, all automation projects implement hybrid frameworks, in the next chapter, we will focus on automating various mobile applications through the user agent approach using the hybrid framework discussed in this chapter.

# 3
# User Agent – automating Mobile Applications with Browsers

In previous chapters, we learned about the concepts of mobile testing automation, and in this chapter we will learn the practical implementation of each of the approaches we have learned till now in greater detail.

In this chapter, we will specifically focus on the automation of a mobile application through the user agent spoofing technique.

This chapter will cover the following topics:

- What are user agents
- User agent simulation or spoofing
- Advantages and disadvantages of this technique
- How to set the mobile test automation lab for this technique
- Sample code explaining the concepts
- Troubleshooting and best practices

# Introduction

When a client's machine accesses a server through a network, the accessing program or application is generally referred to as a **user agent**. So a web browser that is being used to open a website from a server can be termed as a user agent. This process of client-server communication also needs a method through which the server will identify the requesting user agent, so that it can present it with customized content if required. Generally, a string is accompanied by a request from any client program to identify itself to the server, and this string is termed as the **user agent string**. Using the information present in this string, a server identifies the application name, version, operating system, and other necessary details, such as language in which the content may be presented. The process through which servers identify the accessing user agent is referred to as **user agent sniffing**. This was required because during the early days of the Internet, site administrators generally favored one browser over the others and the servers were configured with rules that blocked requests from non-supported browsers completely, or were presented with a highly stripped down version of the website. This saved them the effort of constantly enhancing the code to support newer browsers.

# A user agent string

The user agent string format is specified by *Section 5.5.3* of *HTTP/1.1 Semantics and Content*. A user agent string typically has tokens that carry the specified information with optional comments. Due to the relative popularity and general ubiquitous presence of Internet Explorer after it overtook Netscape Navigator, its user agent string has become the most widely used over the Internet. A typical user agent string, accompanied by an Internet Explorer request, is as follows:

```
<Application Name/Version> (<compatibility flag>;<Version of
browser>;<OS Version>;<MSHTML Version>) <platform> (<platform
details>) <extensions>
```

For example, for IE 9 on Windows 7, the user agent would be as follows:

```
Mozilla/5.0 (compatible; MSIE 9.0; Windows NT 6.1; Trident/5.0)
```

The following is a description of each of these tokens:

| Token name | Token value | Description |
|---|---|---|
| Application name / Version | Mozilla/5.0 | As explained, for historical reasons, even the IE browser identifies itself as a Mozilla browser and indicates its compatibility with the Mozilla rendering engine. |
| Compatibility flag | Compatible | It indicates that the browser is compatible with a set of common features. |
| Version | MSIE 9.0 | This identifies the browser and contains the version number. |
| Platform | Windows NT 6.1 | This identifies the operating system and version as Windows 7. |
| Language | en-us | This indicates the language and locale as English US. |
| MSHTML version | Trident/5.0 | This can be used to determine whether or not the webpage is to be displayed in Compatibility View. |

Apart from the ones mentioned here, it is possible that there are alternate tokens in the user agent string that may be utilized to identify optional features or optional comments sections that may be utilized to attain additional information. For example, bots provide the URL of the originator or a contact e-mail so that webmasters can contact the bot operators, if required.

In these days of pages having **Responsive Design Techniques**, just like other browsers, even mobile devices provide the user agent string to communicate information that is useful for servers to determine the content that needs to be presented to the device. For example, on an iPad, the requesting Safari browser would have a user string, as follows:

```
Mozilla/5.0 (iPad; U; CPU OS 3_2_1 like Mac OS X; en-us)
AppleWebKit/531.21.10 (KHTML, like Gecko) Mobile/7B405
```

The additional tokens that are present, apart from the ones in the general browsers, contain specific information—such as (iPad; U; CPU OS 3_2_1 like Mac OS X; en-us) —that indicates the details of the system in which the browser is running; in this case, iPad. The token AppleWebKit/531.21.10 indicates the platform being used by the browser, and the token (KHTML, like Gecko) indicates the browser platform details. The final token, Mobile/7B405, is used by the browser to indicate specific enhancements that are available directly in the browser or through third-party programs like **LiveMeeting**. Also, it is quite similar to the IE browser portraying itself as Mozilla. Even the Android browser identifies itself as Safari in order to boost its compatibility with mobile-specific websites.

# User agent simulation for mobile

To overcome the problem of user agent sniffing, new browsers were designed to allow for spoofing or simulation of the user agent as a means of bypassing discriminatory server rules. This capability has since then persisted and all modern browsers today have the feature of user agent string simulation as an inbuilt capability. So, using this feature to our advantage, we can simulate the user agent of a mobile device while accessing any mobile web application through any browser. On sniffing the user agent, the server would take the request to be originating from a mobile device and hence would respond with the mobile view. Thus, we would be able to get the mobile web application interface from a desktop browser. The following diagram depicts this process:

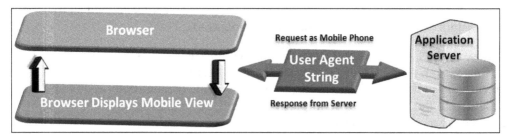

User agent simulation through desktop browsers

# User agent plugin programs

Generally, all mobile browsers come with simple add-ons that enable the overriding of the user agent string. In the following table, some commonly available add-ons and some freeware tools for various modern browsers are provided:

| Browser | User agent switcher tool | Download link |
|---|---|---|
| IE v8 and above | Bayden UA Pick | `http://www.enhanceie.com/ietoys/uapick.asp` |
| IE all versions | Fiddler | `http://www.fiddler2.com/fiddler2/version.asp` |
| Mozilla Firefox | User agent Switcher add-on | `https://addons.mozilla.org/en-US/firefox/addon/user-agent-switcher/` |
| Mozilla Firefox | Modify Headers | `https://addons.mozilla.org/en-US/firefox/addon/modify-headers/` |
| Safari | In built tool | Go to **Preferences** \| **Advanced** \| **Developer menu** |
| Chrome | User agent switcher add-on | Available from Chrome Market |
| Chrome | In Built Developer Tool | Available inbuilt with the latest versions of Chrome |

After these add-ons are installed, the user agents can be manipulated very easily through GUI controls that get added on to the browser. Most automation tools that work with GUI controls can be used to work with the browser GUI controls to replicate the manual process of changing the user agent, as per the required platform. Although this is a cumbersome process, it is very easy to automate.

Some commonly used user agent strings are listed in the following table:

| Model | Manufacturer | OS version | User agent string |
|---|---|---|---|
| iPhone 4S | Apple | iOS 5 | `Mozilla/5.0 (iPhone; CPU iPhone OS 5_0 like Mac OS X) AppleWebKit/534.46 (KHTML, like Gecko) Version/5.1 Mobile/9A334 Safari/7534.48.3` |
| iPhone 5C | Apple | iOS 7 | `Mozilla/5.0 (iPhone; U; CPU iPhone OS 7_0_5 like Mac OS X; xx-xx) AppleWebKit/532.9 (KHTML, like Gecko) Version/4.0.5 Mobile/8A293 Safari/6531.22.7` |
| Nokia Lumia 1020 | Microsoft | Windows 8.0 | `Mozilla/5.0 (compatible; MSIE 10.0; Windows Phone 8.0; Trident/6.0; IEMobile/10.0; ARM; Touch; NOKIA; Lumia 1020)` |
| Galaxy S4 | Samsung | Android OS v4.2 (Jelly Bean) | `Mozilla/5.0 (Linux; U; Android 4.2; xx-xx; GT-I9500 Build/JDQ39) AppleWebKit/534.30 (KHTML, like Gecko) Version/4.0 Mobile Safari/534.30` |
| Galaxy S II | Samsung | Android OS v2.3.4 (Gingerbread) | `Mozilla/5.0 (Linux; U; Android 2.3; xx-xx; SHW-M250S Build/GINGERBREAD) AppleWebKit/525.10 (KHTML, like Gecko) Version/3.0.4 Mobile Safari/523.12.2` |
| EVO Design 4G | HTC | Android OS v2.3.4 (Gingerbread) | `Mozilla/5.0 (Linux; U; Android 2.3; xx-xx; HTC/Shooter_U/1.01.161.1 Build/GRJ22) AppleWebKit/533.1 (KHTML, like Gecko) Version/4.0 Mobile Safari/533.1` |

| Model | Manufacturer | OS version | User agent string |
|-------|--------------|------------|-------------------|
| Atrix 4G | Motorola | Android OS v2.3 (Gingerbread) | `Mozilla/5.0 (Linux; U; Android 2.2; xx-xx; MB860 Build/Blur_Version.4.0.300. MB860.Orange.en.GB Flex/ P009) AppleWebKit/533.1 (KHTML, like Gecko) Version/4.0 Mobile Safari/533.1` |
| Curve 9380 | BlackBerry | BlackBerry OS 7.0 | `Mozilla/5.0 (BlackBerry9380/5.0.0.913 Profile/MIDP-2.1 Configuration/CLDC-1.1 VendorID/104)` |
| 8520 | BlackBerry | BlackBerry Proprietary | `Mozilla/5.0 (BlackBerry; U; BlackBerry8520/4.6.1.259 Profile/MIDP-2.0 Configuration/CLDC-1.1 VendorID/100)` |
| Titan | HTC | Windows 7.5 Mango | `Mozilla/5.0 (compatible; MSIE 6.0; Windows CE; IEMobile 7.11) Sprint: PPC6800` |

In the following section, we will look at the process of setting up user agents with common browsers such as Google Chrome and Mozilla Firefox.

 For newer devices, user agents can be obtained from the manufacturer's support websites or sourced from the Internet. However, caution must be taken when manually generating the user agents, and the user agents should be verified with the development team for accuracy.

# Setting up a mobile test automation lab for user agents

One important consideration while selecting a browser for user-agent-based mobile testing is the rendering engine of the browser and its compatibility with modern mobile presentation software and mobile browsers. For example, one of the most commonly used mobile browser technologies is the **Webkit**. So, it is always advisable to go for modern browsers such as Safari and Chrome while working with mobile web applications that target devices based on the Android and iOS operating systems. And for application testing for devices that carry the Microsoft Windows OS, Internet Explorer would be the most logical choice.

## Setting up the user agent add-on for Mozilla Firefox

The following is the step-by-step process to set up the Mozilla Firefox user agent plugin:

1. To install the user agent add-on in Firefox, go to **Add-ons**, run a search for user agents, and click on **Install** for the **User Agent Switcher 0.7.3** add-on:

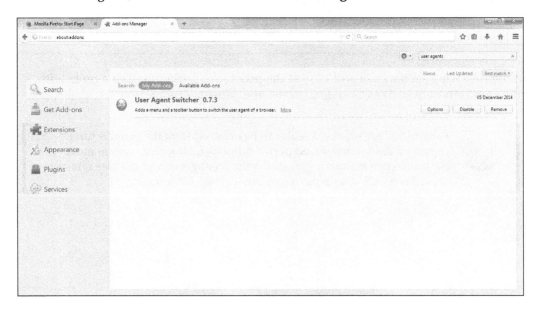

2.  After installing the Firefox user agent add-on from the add-on market, it will be accessible from the **Tools** menu. Since the **Tools** menu is part of the browser, it can be automated with UI automation tools, allowing you to automate the user agent data at the start of the browser launch process.

    The steps required to automate user agents with UI automation tools are shown, as follows:

3.  To add a new user agent string, go to **Tools | Default User Agent | User Agent Switcher | Options | New | New User Agent String**.

    ○  In the **Description** row, add the name of the device for which we are adding the string.

    ○  In the **User Agent** row, add the string of the device.

    ○  In the **Platform** row, provide the OS of the device.

○ In the **Vendor** row, provide the manufacturer of the device:

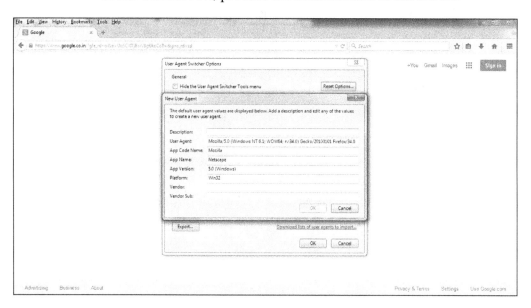

In the following example, let's add a new user agent string for **iPhone 5C**:

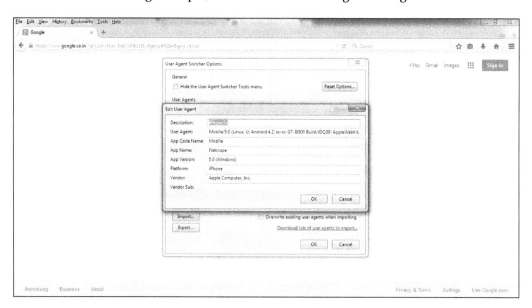

4.  The menu path to manipulate/override the user agent is **Tools | Default User Agent | User Agent Switcher | Options**:

5.  To use any user agent string, go to **Tools | Default User Agent** and select the desired device from the listed options, as shown here:

For example, using iPhone 5C as the default string, the URL www.linkedin.com is displayed as follows:

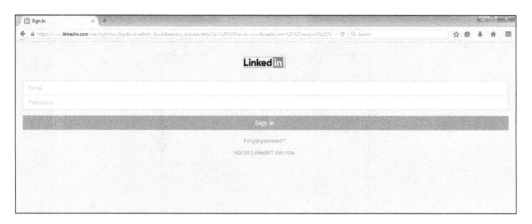

And the same URL on the Mozilla Firefox web browser without the user agent spoofed or overridden to mobile is displayed as follows:

# Setting up a user agent add-on for Google Chrome

For Google Chrome, although there are plenty of user agent switcher add-ons available, we have taken an example of one of the most highly rated ones. It is called the **Chrome UA Spoofer**.

The following is the step-by-step process to set up the Chrome user agent plugin:

1. To install a user agent in Chrome, go to **Chrome Web Store** and run a search for the **Chrome UA Spoofer** add-on.

2. After installing the Chrome user agent add-on from the add-on market, its icon will be displayed near the search bar, as shown in following screenshot. Generally, it comes with a list of preloaded default user agent strings that are commonly used. However, these user agents should also be verified with the application development team before using them as is:

3. To add a new user agent string, right-click on the **User Agent Switcher** icon and click on **Options**. Then follow these steps:

   ° In the **New User-agent name** column, add the name of the device for which we are adding the string.

   ° In the **New User-Agent String** column, add the string of the device.

   ° In the **Group** column, enter the operating system of the device.

   ° From the **Append** dropdown, select the **Replace** option.

   ° In the **Indicator Flag** column, enter the short name of the device for identification.

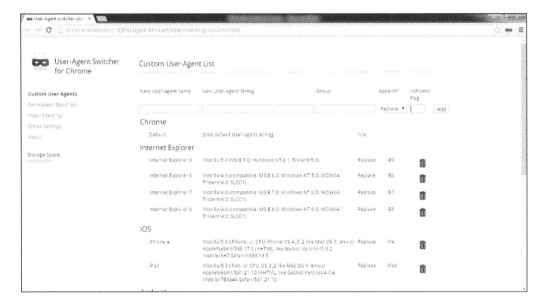

4. To delete the user agent, right-click on the **User Agent Switcher** icon and go to **Options**.

5. To use any user agent string, click on the **User Agent Switcher** icon and then, from the appropriate group, select the desired device as shown in the following image. After we have done the setting, the added user agents are readily available in the options under **Chrome User Agent**. Along with this tool, Chrome also provides an in-built user agent switcher option within the **Developer Tools** option, as shown in the following screenshot:

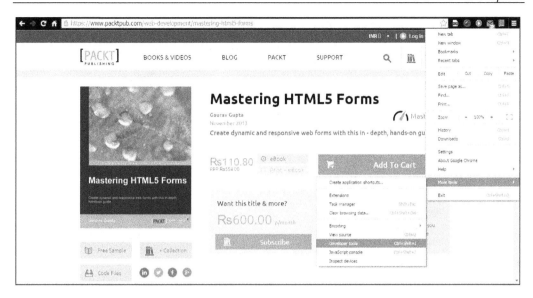

For example, using iPhone 5C as the default string, the URL www.linkedin.com is displayed as follows:

And the same URL on a web browser is displayed as follows:

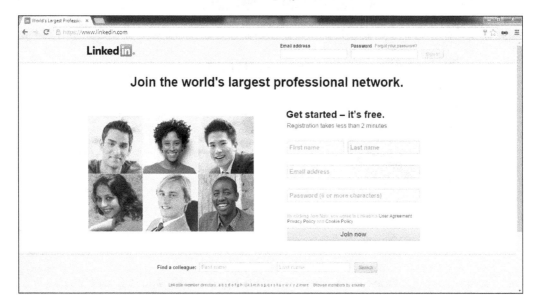

With the **Chrome Developer** option, the website can be viewed as it would appear in a mobile device:

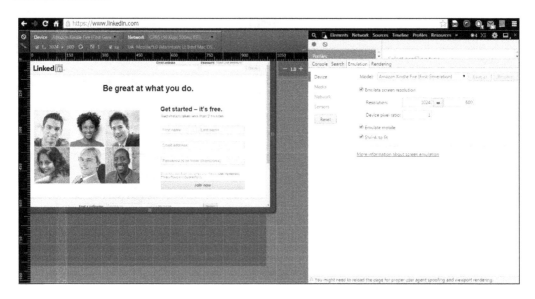

Overall, the most suitable browser for user-agent-simulation-based functional and automation testing of mobile web application is Google Chrome, for the following reasons:

- It uses the **Webkit** rendering engine (`https://www.webkit.org/`), which provides the closest approximation of the mobile presentation layer used with modern mobile operating systems, such as Google Android and Apple iOS. Due to this, the mobile-specific controls are displayed more accurately on Google Chrome.

- Out of the two widely used browsers that use the Webkit rendering engine, Google Chrome and Safari, Chrome is the most widely supported browser with open source and commonly used automation tools, such as Selenium and HP UFT.

- It provides functionality to programmatically override the user agent with a simple command line. Moreover, its capabilities can be manipulated with Selenium WebDriver, making it easy to automate with easy-to-use commands.

So it is advisable to set up the desired user agent strings for the mobile platforms and device versions in scope on various machines and then trigger the execution using the predefined set of user agents.

# Advantages of this approach

As explained in *Chapter1*, *Ensuring Five-star Rating in the MarketPlace*, the user agent approach to mobile automation provides a very quick and easy to set up process for mobile automation. Here's why:

- **Support for all platforms**: With the user agent approach, by simple manipulation of the user agent string, the automation scripts can be executed over any given platform and browser combination. In this way, a very detailed and thorough browser validation suite can be developed, which results in a prompt **Return on Investment (RoI)**.

- **Quick implementation**: This approach requires no additional setup, other than the browser user agent add-ons, and in some cases even that can be bypassed by triggering the user agent manipulation programmatically. Thus it is very easy to implement automation solutions and the setup may be done with minimal effort.

- **Cost effectiveness**: Of all the automation approaches, the user-agent-based mobile automation is the most cost-effective way as it requires no additional hardware or software, other than what is normally available for any functional automation project. Since browsers and their add-ons for user agent manipulation are freeware, it doesn't entail any additional cost and nor is the effort too high, which brings down the costs. Also, the automation of such scripts is also almost similar to traditional automation projects, meaning there are no additional training and licensing costs.

- **Can leverage freeware like Selenium for automation**: Since this approach requires the automation of browsers, it can be done with open source tools such as **Selenium** and **WATiR**, which are open source and provide a robust support framework for automating browsers. This helps simplify the implementation and maintenance of the automation suite.

- **Leveraging the existing automation setup**: If an organization that is just starting their mobile automation already have an existing regression suite for their web applications, then it can be easily modified to support user-agent-based automation. Thus, it can help reduce the setup time and effort, as well as remove any additional training requirements to help bring the automation team up to speed for mobile automation. Any functional automation framework can be made suitable for user-agent-based automation, as explained in the previous section, by creating a mechanism to programmatically control the user agent from a central database or file. Thus, the user agent approach serves the purpose of enabling any existing functional automation suite to be used for mobile automation.

# Limitations of this approach

As discussed briefly in *Chapter 1, Ensuring Five-star Rating in the MarketPlace*, in the user agent approach, the disadvantages stem primarily from the fact that it is a purely simulation-based approach, which is limited to only mobile web applications.

The following are the limitations that must be thoroughly understood and gradually addressed with a mixed approach towards the automation of mobile applications:

- **Limited to only mobile web applications**: The user agent approach cannot be extended to native mobile applications; although hybrid applications can, to an extent, be validated with this approach, it cannot be tested from within the application-specific shell. Hence, in essence, this approach can reliably only be used for mobile web application browser certification testing.

- **Device-specific issues cannot be captured**: Many mobile web applications utilize device-specific features such as geolocation coordinates, which cannot be easily replicated with this approach. Although some browsers support supplying geolocation coordinates now, it is not a very reliable approach for testing and is limited to manual testing only, as automation of such features is not allowed (for security concerns). Thus, any scenarios where such device-specific features need to be tested cannot be recreated with this approach.

- **Cannot be used for UI-related test cases**: Mobile devices vary particularly in their UI layer, due to the form factor and operating-system-specific features, such as the screen resolution and pixel density that might impact the application user interface. These features cannot be reliably tested with this approach and it is thus generally recommended to utilize the user-agent-based mobile automation approach only for functional automation, after a thorough round of UI testing has already taken place. Also, for verification, any UI-specific issues must preferably be replicated on a real device, because it has been observed that some controls, such as pick lists, do not behave in the same fashion, even with the user agent mode switched on, and work only on real devices. Also, user interactions that are common on mobile devices, particularly related to gestures and touch screens are not supported and thus any test scenarios that require extensive validation of such controls must preferably be done on real devices.

These limitations thus imply that although the user agent approach has its advantages, it cannot be solely relied upon for a comprehensive automation suite. It should ideally be used in conjunction with another approach to address these shortcomings adequately. We will discuss more of this aspect in the *Chapter 8, Delivering Customer Delight*, which is about optimization and ensuring customer satisfaction.

# The Hybrid framework implementation for the user agent

Until now, in this chapter, we have understood the manual process of how user agent manipulation works and how simulation can be used for mobile testing. The following steps will help us understand how to set up a mobile automation framework that can be used for unmonitored execution of mobile web applications without the use of actual devices:

1. Identify the various platforms for which the **Application Under Test (AUT)** needs to be validated.

2.  Identify the user agent switcher tool that corresponds to the selected browser to be used for testing, and set it up as per the process explained in the previous section.

3.  Identify the user agent string for all the platforms that you need and set them up in the user agent switcher tool.

4.  Set up the user agent string repository in an external file, such as an Excel sheet or a Database, that may be periodically updated in order to execute the scripts over newer versions of operating systems and devices.

5.  Set up the functional framework, which ideally may be a hybrid framework, and include the user agent lookup components (user agent select and user agent set).

In the following schematic diagram, we have shown a conceptual implementation of the hybrid automation framework for the user agent approach. The **Execution Manager** is the control mechanism that instructs the driver script to execute selective test cases. This may be a simple Excel sheet or a text file with entries of the test case names that need to be executed, along with details of the device on which the execution, needs to be triggered. The driver script then connects to a logical code block that we have referred to as the user agent engine. There are three major components in this section:

1.  **User Agent selector**: This component reads the user agent input file to provide other components that are the required input string. It first connects to the execution manager or an external configuration file that provides the ID of the user agent that needs to be used for a particular test script or flow. Then it connects to the database or the file that contains the actual user agent string information, along with any other relevant data, and provides it as an output for other components to consume.

2.  **User Agent setter**: This sets up the provided user agent through any of the available techniques. The simplest technique is to use the user agent manipulation add-on tools mentioned in the previous section through the manipulation of GUI controls. Another technique is to utilize the programmatic setting of the user agent string through options such as command-line variables, while launching the required browser.

3.  **User Agent switcher**: This component is used to switch the user agent during the execution by either selecting a preexisting user agent string or by using a predefined profile, created manually either at the onset of the project or during the project setup.

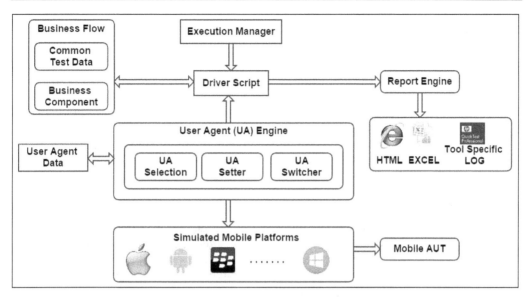

Hybrid framework for user-agent-simulation-based automation

# The UFT code snippet

The following is the VB-Script-based UFT-specific implementation of the user agent engine portion of the hybrid framework, explained in the previous section. As shown in the preceding figure, the major components are **UserAgentSelector**, **UserAgentSetter**, and **UserAgentSwitcher**. The logical blocks for **UserAgentSwitcher** and **UserAgentSetter** have been merged into a single component in the following code, and is named **UserAgentSwticher**.

Since the following code is lengthy, for better readability and understanding, we will be including the code explanation within the snippet itself:

```
Function UserAgentSwitcher
Dim WshShell, i, UA, AppURL
```

The calling function will fetch user agent strings from the mobile input data sheet, as per the test case:

```
UA=UserAgentSelector
Systemutil.Run "chrome.exe"," -user-agent="&Chr(34)&UA&chr(34)
Window("Google Chrome").Activate
'Fetching URL
AppURL = DataTable("ApplicationURL",dtGlobalSheet)
Window("Google Chrome").WinObject("Chrome_seturl").Type AppURL
Window("Google Chrome").WinObject("Chrome_seturl").Type micReturn
```

```
Browser("Browser").Sync
End Function
'###################################################################
'Function Name: UserAgentSelector
'Description   : This function will fetch device information from
mobile input data sheet as per execution manager setting
'###################################################################
Function UserAgentSelector
Dim objExcel, objWorkbook, objSheet, column_count, row_count,
UA_ID, UA_String, i, User
Dim UserAgentType, UserAgentModel, UserAgentOSVersion,
UserAgentManufacturer
```

By using the command-line flag --user-agent, the Chrome browser can be launched with the user agent string as a parameter that sets up the user agent with the launched session, thereby enabling the emulation. For Mozilla Firefox, the same can be achieved by using the **about:config** option.

The device user agent string should be indicated against each test case as an ID that can be fetched from **External Input Test Data Sheet**. For UFT, this can also be from the inbuilt data table that can double up as the **Execution Manager** sheet.

```
User = DataTable("UserAgentString",dtGlobalSheet)
Set objExcel = CreateObject("Excel.Application")
'Getting the path of Mobile Input Test Data Sheet using Relative
Path
Set objWorkbook = objExcel.WorkBooks.Open
((Environment.Value("RelativePath"))
&"\Datatables\UserAgentData.xls")
Set objSheet = objWorkbook.Worksheets("UserAgent")
row_count= objSheet.usedrange.rows.count
For i=2 to row_count
UA_ID=objSheet.cells(i,1).value
```

Some additional values can be stored in the user agent string information sheet, which can then be used for more effective reporting of the execution logs:

```
If (Trim(User) = Trim(UA_ID )) Then
UA_String=objSheet.cells(i,5).value
UserAgentType=objSheet.cells(i,6).value
UserAgentModel=objSheet.cells(i,2).value
UserAgentOSVersion=objSheet.cells(i,4).value
UserAgentManufacturer=objSheet.cells(i,3).value
Exit For
```

```
End If
Next
UserAgentSelector = UA_String
Environment.Value("UAType") = UserAgentType
Environment.Value("UAModel") = UserAgentModel
Environment.Value("UAOSVersiont") = UserAgentOSVersion
Environment.Value("UAManufacturer") = UserAgentManufacturer
'Closing and releasing the memory allocated to excel object
objWorkbook.Close
Set objExcel = Nothing
Set objWorkbook = Nothing
Set objSheet = Nothing
End Function
```

# The Selenium code snippet

In the following code snippet, we have explained the Selenium-based approach to set up the user agent. The major components provided in this implementation are the `UserAgentSetter` class, which sets the user agent as per the requirement, and the `UserAgentSelector` class, which reads the input file to find which user agent is to be used against which test case. The code is provided here with explanatory comments:

```
import java.io.BufferedReader;
import java.io.FileReader;
import org.openqa.selenium.WebDriver;
import org.openqa.selenium.chrome.ChromeDriver;
import org.openqa.selenium.chrome.ChromeOptions;
import org.openqa.selenium.firefox.FirefoxDriver;
import org.openqa.selenium.firefox.FirefoxProfile;
import org.openqa.selenium.remote.CapabilityType;
import org.openqa.selenium.remote.DesiredCapabilities;
import SupportLibraries.Util;

public class UserAgentSetter {
  public static WebDriver userAgentSetter(String browser,String
  strUA) {
//checking which browser type is to be used for execution
    WebDriver driver=null;
    if(browser.equals("firefox")) {
      FirefoxProfile profile = new FirefoxProfile();
      if(strUA!=null) {
```

While initializing the Firefox profile, a preference is set that allows us to override the existing user agent with the one provided.

```
        profile.setPreference("general.useragent.override",strUA);
    }
    DesiredCapabilities capability =
    DesiredCapabilities.firefox();
    capability.setCapability(FirefoxDriver.PROFILE, profile);
    capability.setCapability(CapabilityType.ACCEPT_SSL_CERTS,
    true);
    capability.setCapability(CapabilityType.
    SUPPORTS_APPLICATION_CACHE, true);
    capability.setCapability(CapabilityType.
    SUPPORTS_BROWSER_CONNECTION, true);
    capability.setCapability(CapabilityType.
    SUPPORTS_LOCATION_CONTEXT, true);
    capability.setCapability(FirefoxDriver.PROFILE, profile);
    driver = new FirefoxDriver(capability);
}
else if(browser.equals("chrome")) {
    System.setProperty("webdriver.chrome.driver",
    "c:\\chromedriver.exe");
    ChromeOptions options = new ChromeOptions();
    if(strUA!=null) {
```

For the Chrome browser, an argument with the flag user-agent , along with the user agent string to be used, is provided to simulate the mobile platform:

```
        options.addArguments("user-agent="+strUA);
    }
    options.addArguments("start-maximized");
    DesiredCapabilities capability =
    DesiredCapabilities.chrome();
    capability.setCapability(CapabilityType.ACCEPT_SSL_CERTS,
    true);
    capability.setCapability(CapabilityType.
    SUPPORTS_APPLICATION_CACHE, true);
     capability.setCapability(CapabilityType.
     SUPPORTS_BROWSER_CONNECTION, true);
     capability.setCapability(CapabilityType.
     SUPPORTS_LOCATION_CONTEXT, true);
     capability.setCapability(ChromeOptions.CAPABILITY,
     options);
     driver = new ChromeDriver(capability);
```

```
    }
    return driver;
}
```

 With the Chrome Web Driver, an experimental option is also available, which can be used to access mobile emulation from Developer options as follows:

```
chrome_options.add_experimental_
option("mobileEmulation"
, mobile_emulation)
```

The userAgentASelector() function reads the Config.ini file provided under the relative path, containing the user agent string information:

```
public static String userAgentSelector() {
  String strUA=null;
  BufferedReader br = null;
  String strLine = "";
  String strKeyval = "";
  Boolean boolFound = false;
  String path = "\\config.ini";
  Try {
    br = new BufferedReader(new FileReader(path));
    String delimeter = "=";
    while ((strLine = br.readLine())!=null) {
      if (strLine.trim().length()>0) {
        String[] pairs = strLine.split(delimeter);
        if (pairs[0].trim().equals("UAString")) {
          strKeyval = pairs[1];
          strUA = strKeyval;
          boolFound = true;
        }
      }
    }
  }
  catch(Exception e) {
    e.printStackTrace();
  }
  return strUA;
}
public static void main(String args[]) {
  WebDriver driver = userAgentSetter("chrome",
  userAgentSelector());
  driver.get("http://www.google.com");
```

```
        driver.quit();
    }
}
```

# Troubleshooting and best practices

The following are some industry best practices:

- An important aspect to consider with the user agent approach is that while switching user agents, a new session of the browser should be started every time in order to ensure that cookies are cleared, and the server is able to send the response as per the new user agent string.

- Another aspect is to periodically clear all temporary files of the browser and also to clear its cache while running automation scripts. The following is a code snippet for VB Script that can be used with tools such as UFT, which do not have inbuilt constructs for deleting browser cookies or whose inbuilt cookie deletion functions do not work properly:

```
Dim objFileSystem, objWshShell, objoFolder, oFile
'Deleting temporary folders:
Set objFileSystem =
CreateObject("Scripting.FileSystemObject")
'Create Window Shell object
Set objWshShell = CreateObject("WScript.Shell")
'get Folderpath under which file present
Set objoFolder =
objFileSystem.GetFolder(objWshShell.ExpandEnvironmentString
s("C:\Users\%username%\AppData\Local\Temp"))
'Get count the files under the folder and loop run for
delete the files
For Each oFile In objoFolder.files
objFileSystem.DeleteFile oFile
Err.clear
Next
'Delete the subfolders under the temp folder
For Each oSubFolder In objoFolder.SubFolders
objFileSystem.DeleteFolder oSubFolder
Err.clear
Next
set objFileSystem = nothing
set objWshShell = nothing
set objoFolder = nothing
Err.clear
```

The following code can be used to clear the temporary files, history, and cookies in your Chrome browser before initializing:

```
Set WshShell = CreateObject("WScript.Shell")
WshShell.SendKeys "+^{DELETE}"
For i = 1 to 8
WshShell.SendKeys "{TAB}" wait(1)
Next
WshShell.SendKeys "{ENTER}"
Set WshShell = Nothing
'Closing Chrome after deleting temporary file, history and
cookies
SystemUtil.CloseProcessByName("chrome.exe")
```

- Sometimes, we might observe that the layout of objects is not proper, so for GUI testing, this approach is not very reliable and should be followed up with at least one round of a GUI testing phase.

- It is important to design the business components of an automation suite to handle the differences between objects and layout across different device types, form factors, and operating systems. This is crucial because with this way of implementation, the keyword part of the framework remains independent of the device-based variations in the application, resulting in only unique business components and test cases for all application functionality. This helps to reduce the framework implementation complexity and it makes the maintenance of the framework easier.

# Summary

In this chapter, we learned about the user-agent-simulation-based automation of mobile web applications. This automation approach is one of the quickest to set up and easiest to implement. However, since it is pure simulation, care must be taken to not overly rely on this approach, even for pure mobile web applications, as there a few scenarios where only a real device can be utilized, such as location-based testing and device interaction features such as swiping and scrolling. With a robust implementation of the framework, this approach can help you set up a comprehensive regression test suite that can be executed reliably over a range of devices and operating systems.

In the next chapter, we will look into the use of emulators and simulators and various techniques to automate emulated devices.

# 4
# Emulators and Simulators – the Automation of Emulated Devices

An **emulator** is software that mimics the complete behavior of any other software and has the same or similar code to execute, whereas a **simulator** just replicates the behavior in an external sense. That is why we have flight simulators and not flight emulators. However, this distinction is more academic than practical since Google provides Android emulators, whereas Apple prefers to call their iOS ones simulators, even when both of them perform roughly the exact same function. For better readability, we'll refer to all iOS ones as simulators, and for rest of them, we will use the term emulators.

When generally speaking, though, we prefer to call this technique emulator-based, since they are closer in approximated behavior to real devices than the user agent approach, which we prefer to call simulated mobile browsers.

In this chapter, we will cover the following topics:

- Introduction to mobile emulators
- How to set up mobile emulators
- Commonly used automation tools and support frameworks
- Implementing a mobile test automation framework for this technique
- Advantages and limitations of this technique
- Troubleshooting and best practices

# Introduction to mobile emulators

As discussed in *Chapter 1, Ensuring Five-star Rating in the MarketPlace*, a device emulator is a desktop application that emulates both the mobile device hardware and its operating systems, thus allowing the testing of applications a lesser degree of tolerance and better accuracy. A simulator may be created by the device manufacturer or by some other company that offers a virtual environment; please note, however, that simulators are less accurate; than emulator programs.

In most discussions, though, the terms emulator and simulator are used interchangeably. Emulators can be understood as virtual device hardware that operates on a PC or laptop. Emulator software can be obtained from platform and device makers. In some cases, emulators come bundled with the **Software Development Kit** (**SDK**); for example, Google's Android SDK has a mobile emulator. While, in other cases, it can be a separate program altogether, such as Apple's iOS simulator, which can be used to run iOS on a Mac machine. It is important to note that the iOS simulator also comes bundled with XCode and there are tools like xPadian's **iPadian**, which is a third-party iOS simulator.

In the following table, we have provided some commonly available emulator software and the links to download them:

| Emulator | Provider | Download URL |
|---|---|---|
| Android | Google | `http://developer.android.com/sdk/` |
| iOS | Apple | `https://developer.apple.com/xcode/ide/` |
| iOS | xPadian | `www.xpadian.com` |
| Windows Phone | Microsoft | `http://dev.windows.com/en-us/develop/download-phone-sdk` |
| Blackberry | RIM | `http://us.blackberry.com/sites/developers/resources/simulators.html` |
| Android | Samsung | `http://developer.samsung.com/sdk-and-tools` |
| Android | Motorola | `http://source.android.com/source/building-running.html#emulate-an-android-device` |
| Firefox OS Simulator | Mozilla | `https://ftp.mozilla.org/pub/mozilla.org/labs/fxos-simulator/` |

Emulator programs were created initially as a means to allow mobile application developers to do quick and dirty validations of their applications during the design phase. So, they were designed to be used by developers in the early stages of the software development phase. But their potential for functional testing was quickly realized by testing teams, who realized that the majority of testing could be carried out on these emulators. Also, since one is still working on a PC environment, taking screenshots of errors and capturing logs are also relatively hassle-free than when working with a real device. So, in essence, emulators have allowed stakeholders to bring down the cost of projects that require large-scale validation on multiple devices, because with emulators in place, a project manager doesn't have to worry about arranging all devices for all developers and testers.

# Setting up a mobile emulators for automation

In this section, we will take up examples of Android and iOS emulators in order to understand the setup required on a Windows machine, so that we can test the mobile applications on different platforms, without using a real device.

We can follow a similar process on a Linux or Mac machine and even for other emulator software with minor changes to commands.

## The Android emulator setup

The following is the step-by-step process required to set up the Android emulator on a Windows machine:

1. Download the Android SDK or the complete Android Studio from `http://developer.android.com/sdk/index.html`. Note that you need to download according to the required machine configuration, such as 32- or 64-bit.

2.  Now download and install **Android SDK Tool** and **Android SDK Platform-tools**:

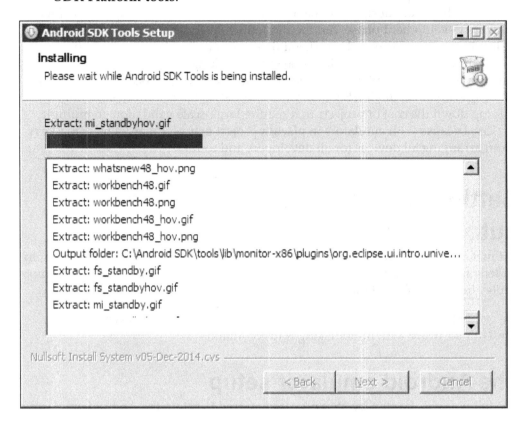

3.  Now, to set the `Path` environment variable for the Android SDK, follow these steps:

    1.  Open the **System** properties window by right-clicking on **My Computer** , and then on **properties**.

    2.  Click on the **Advanced system settings** link and go to the **Advanced** tab.

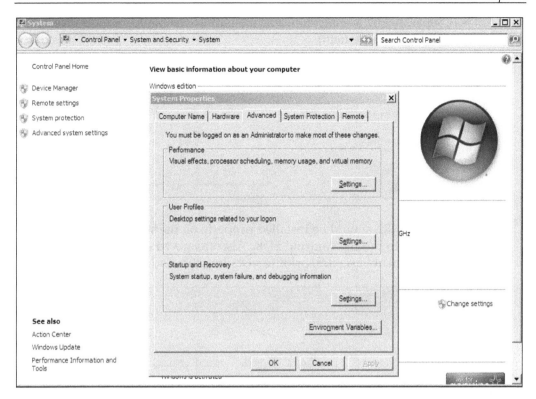

3.  Click on **Environment Variables** and define a new variable ANDROID_
    HOME , with the path to the SDK directory as **Variable value**:

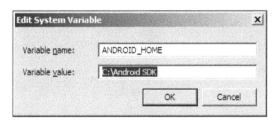

4.  Under system variables, locate the Path variable and add the path to
    the tools directory as **Variable value**:

Then, separated by a semicolon, add the path to the platform tools directory to the **Variable value** field:

Java JDK and JRE should be installed properly on the system, as they are prerequisites to run the Android SDK. Also ensure you set up the proper Java environment variables, prior to installing the Android SDK:

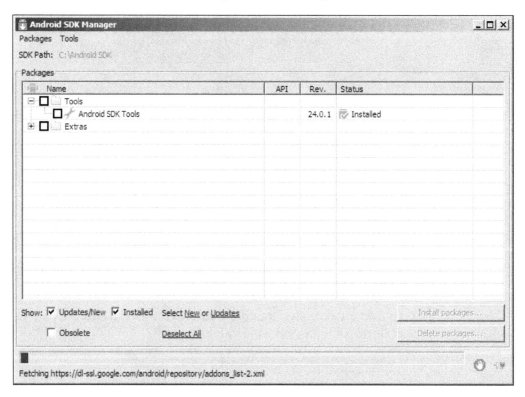

4.  After the installation is completed, in order to verify that Android SDK is installed properly, we use the ADB command.

5. The ADB command is `android debug bridge`, which can be used to gather debug-level information while you are testing with an **Android Virtual Device (AVD)**. It will display some basic information, as follows:

```
Administrator: C:\Windows\system32\cmd.exe                          _ 8 X

C:\>android debug bridge
Usage: java [-options] class [args...]
           (to execute a class)
   or  java [-options] -jar jarfile [args...]
           (to execute a jar file)
where options include:
    -d32          use a 32-bit data model if available
    -d64          use a 64-bit data model if available
    -client       to select the "client" VM
    -server       to select the "server" VM
                  The default VM is client.

    -cp <class search path of directories and zip/jar files>
    -classpath <class search path of directories and zip/jar files>
                  A ; separated list of directories, JAR archives,
                  and ZIP archives to search for class files.
    -D<name>=<value>
                  set a system property
    -verbose:[class|gc|jni]
                  enable verbose output
    -version      print product version and exit
    -version:<value>
                  require the specified version to run
    -showversion  print product version and continue
    -jre-restrict-search | -no-jre-restrict-search
                  include/exclude user private JREs in the version search
    -? -help      print this help message
    -X            print help on non-standard options
    -ea[:<packagename>...|:<classname>]
    -enableassertions[:<packagename>...|:<classname>]
                  enable assertions with specified granularity
    -da[:<packagename>...|:<classname>]
    -disableassertions[:<packagename>...|:<classname>]
                  disable assertions with specified granularity
    -esa | -enablesystemassertions
                  enable system assertions
    -dsa | -disablesystemassertions
                  disable system assertions
    -agentlib:<libname>[=<options>]
                  load native agent library <libname>, e.g. -agentlib:hprof
                  see also, -agentlib:jdwp=help and -agentlib:hprof=help
    -agentpath:<pathname>[=<options>]
                  load native agent library by full pathname
    -javaagent:<jarpath>[=<options>]
                  load Java programming language agent, see java.lang.instrument

    -splash:<imagepath>
                  show splash screen with specified image
See http://www.oracle.com/technetwork/java/javase/documentation/index.html for m
ore details.

C:\>
```

6. To create the Android Virtual Device, one final data point is needed about the target list of devices installed. For this, run the following command:

```
prompt>android list targets
```

This will display information about the available Android targets in the following manner:

```
Administrator: C:\Windows\system32\cmd.exe                    _ |□| ×|
id: 1 or "android-18"
        Name: Android 4.3.1
        Type: Platform
        API level: 18
        Revision: 3
        Skins: HUGA, QUGA, WQUGA400, WQUGA432, WSUGA, WUGA800 (default), WUGA854, W
XGA720, WXGA800, WXGA800-7in
        Tag/ABIs : default/armeabi-v7a

id: 2 or "android-21"
        Name: Android 5.0.1
        Type: Platform
        API level: 21
        Revision: 2
        Skins: HUGA, QUGA, WQUGA400, WQUGA432, WSUGA, WUGA800 (default), WUGA854, W
XGA720, WXGA800, WXGA800-7in
        Tag/ABIs : default/armeabi-v7a, default/x86_64

id: 3 or "Google Inc.:Google APIs:18"
        Name: Google APIs
        Type: Add-On
        Vendor: Google Inc.
        Revision: 3
        Description: Android + Google APIs
        Based on Android 4.3.1 (API level 18)
```

 More system images and API levels can be installed by downloading the corresponding versions from http://developer.android. com/sdk/installing/adding-packages.html.

7. Now, navigate to the tools directory and create an Android Virtual Device, using the following command. An Android Virtual Device replicates a real device down to the kernel level and recreates a device with some basic preinstalled programs, such as the caller, take a look at the following commands:

```
prompt>cd ~/android_sdk/tools/

prompt>android create avd -n testing_android -t 1 -c 250M
```

In this command, -n is used to specify the name of the AVD, -t is used to specify the platform target API number, as displayed with the `android list targets` command, and -c specifies the SD card storage space of the emulated device.

This process can also be done manually by using the AVD manager, which is installed along with the Android SDK:

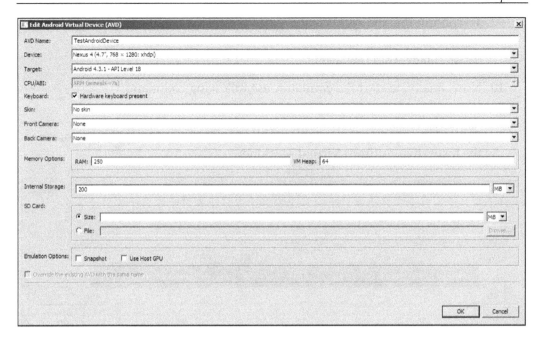

8. Now, after you have entered the required details in the window to create the AVD, click on the **OK** button. A window with the details will be displayed, which is as follows:

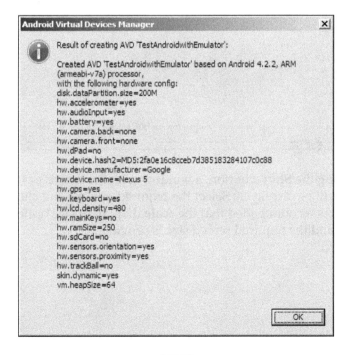

9.  To confirm that the AVD creation was successful, run the following command to list the created AVDs:

    ```
    prompt>android list avd
    ```

10. Now, finally, to start the emulator, use following command:

    ```
    prompt>emulator -avd TestAndroidDevice
    ```

    Optionally, we can use the **Start...** button in the AVD Manager window for the respective Android Virtual Machine:

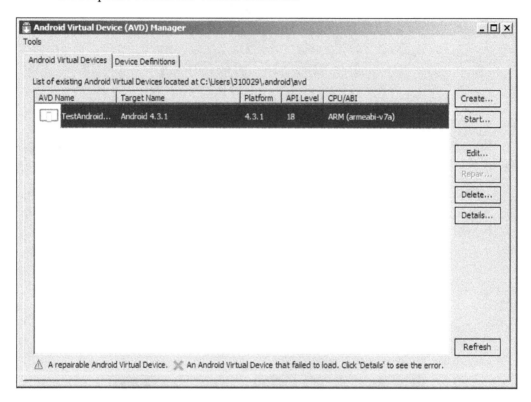

11. On clicking the **Start...** button, a window with the details of the emulator created will be displayed. Select the required options and click on the **Launch** button. It is recommended that the **scale display to size** option is kept checked and the required screen size is provided:

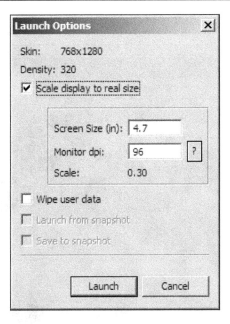

12. Lastly, by clicking on the **Launch** button, the emulator window will be launched:

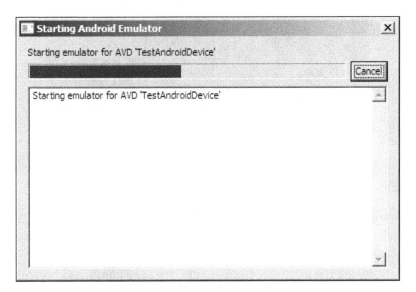

13. The emulator window will be launched with the default applications that are provided with Android.

14. The following screenshot is of the emulator displayed after launch:

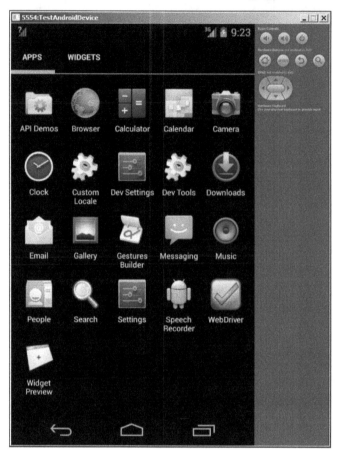

 To install the application under test (if it is not a web application), separate steps need to be performed, as explained in the following steps.

15. To verify that the creation and launch were done properly, after the emulator window is launched, run the `adb devices` command to list the emulator that was just created.

16. If you started the emulator but the `adb devices` command returns no result, try running the following commands in sequence:

```
prompt> adb kill-server
prompt> adb start-server
```

To get the list of all Android commands, use the `help` command as follows:

```
prompt>adb -help
```

# Automating the Android emulator with Selenium WebDriver

In this section, we are going to automate Android Emulator with Selenium WebDriver. The following is the step-by-step process:

1. Every emulator is automatically assigned a serial number that can be used to identify it on the system. To retrieve the serial ID of the AVD currently running on the system, run the following command:

   **prompt>adb devices**

2. Download the Android server from the following location and save it in the `platform-tools` directory, under the `android-sdk` directory:

3. `http://selenium-release.storage.googleapis.com/index.html`

4. To install the application, run:

   **prompt>adb -s <serialId> -e install -r android-server.apk**

5. Start the Android WebDriver application by running this command:

   **prompt>adb -s <serialId> shell am start -a android.intent.action. MAIN -n org.openqa.selenium.android.app/.MainActivity**

   Take a look at the following screenshot:

```
Administrator: C:\Windows\system32\cmd.exe                              _ □ X

C:\>adb -s emulator-5554 shell am start -a android.intent.action.MAIN -n org.ope
nqa.selenium.android.app/.MainActivity
Starting: Intent { act=android.intent.action.MAIN cmp=org.openqa.selenium.androi
d.app/.MainActivity }

C:\>_
```

6. Now we need to set up port forwarding, in order to forward traffic from the host machine to the emulator. Enter the following in to the terminal:

```
prompt>adb -s <serialId> forward tcp:8080 tcp:8080
```

# Code snippet

After we have properly set up the Android emulator with port forwarding, in the following sample script, we will see how to use Selenium AndroidDriver to open the browser on the emulated Android device and perform the following steps:

1. Open www.google.com.

2. Enter the search string Packt Mobile Automation.

3. Verify that Packt is contained in the resulting page.

In the following program, we will import the Selenium AndroidDriver in order to interact with the application controls:

```
import org.openqa.selenium.By;
import org.openqa.selenium.WebDriver;
import org.openqa.selenium.android.AndroidDriver;
import org.testng.Assert;
import org.testng.annotations.AfterClass;
import org.testng.annotations.BeforeClass;
import org.testng.annotations.Test;

public class EmulatedAndroidMobile {
  private static WebDriver driver;
  @BeforeClass
  public void setUp() throws Exception {
    driver = new AndroidDriver();
    driver.get("http://www.google.com");
  }
  @AfterClass
  public void tearDown() {
    driver.close();
    driver.quit();
  }
  @Test
  public void searchGoogle() {
    Thread.sleep(1500);
    driver.get("http://www.google.com");
    // Find the text input element by its name
```

```
        WebElement element = driver.findElement(By.name("q"));
        // Enter search string
        element.sendKeys("Packt Mobile Automation");
        // Now submit the search
        element.submit();
        String searchHeader = driver.findElement(By.cssSelector
        ("H1")).getText().toLowerCase();
        Assert.assertTrue(searchHeader.contains("packt"));
    }
}
```

As you can see, we can utilize Selenium AndroidDriver to open a web browser in an emulated Android device and perform tests directly. However, for Hybrid and Native applications, we need to use alternative solutions. We will discuss the **Appium** automation solution for this in subsequent sections. But before that, let's first understand how to set up an iOS simulator on a Mac desktop PC.

# The iOS simulator setup

In this section, we will learn the setup required to execute the basic web application scripts by using the Selenium iPhone driver in a virtual iOS device. The iOS simulator is bundled with Xcode and Native SDK on Mac, which is available only for devices that carry the Mac operating system.

Follow these steps to set up the iOS simulator and run scripts, using Selenium iPhone driver:

1. Download Xcode from `http://developer.apple.com/iphone or from the built-in App store` and install it on your Mac machine.

2. The iOS simulator comes with only a limited set of capabilities. Some of the prominent shortcomings are:

   ° Features based on the device's hardware, such as accelerometer and gyroscope, cannot be emulated.

   ° URI schemes cannot be emulated. For example, when you click on a number in some text on a phone, the dialer appears with the number, which can then be used to make a call. Such features cannot be emulated with iOS simulator.

   ° It doesn't allow access to other browsers, such as Opera Mini, so these browsers cannot be tested on a simulated iOS device.

   ° It does not have access to any **App Store**, such as iTunes, and thus a manual build process needs to be followed in order to install applications.

3. Since we need Selenium iWeb Driver to execute the scripts, we also need to check out and build it along with the simulator. For this, run the following command on the terminal window to check out while running the SVN tool:

```
prompt>svn checkout http://selenium.googlecode.com/svn/trunk/
selenium-read-only
```

4. In the Xcode IDE, open the following project:

```
selenium-read-only/iphone/iWebDriver.xcodeproj
```

5. Now, set the build configuration to the latest version of iPhone simulator by double-clicking on the project and using the drop-down list that opens.

6. Now, click on the **Play** button to emulate an iPhone.

7. It would appear with the iWebDriver application installed.

8. Lastly, this emulated device can now be used to execute scripts on the Safari browser by using the iPhoneDriver.

## Code snippet

The following is a code snippet that can be used to trigger the same test case as we ran on the Android emulator:

1. Open www.google.com.

2. Enter the search string Packt Mobile Automation.

3. Verify that Packt is contained in the resulting page.

In this program, we have imported the Selenium IPhoneDriver to interact with the emulator browser:

```
import org.openqa.selenium.By;
import org.openqa.selenium.WebDriver;
import org.openqa.selenium.iphone.IPhoneDriver;
import org.testng.Assert;
import org.testng.annotations.AfterClass;
import org.testng.annotations.BeforeClass;
import org.testng.annotations.Test;

public class SimulatediPhone {
  private static WebDriver driver;
  @BeforeClass
  public void setUp() throws Exception {
    driver = new IPhoneDriver();
    driver.get("http://www.google.com");
```

```
    }

    @AfterClass
    public void tearDown() {
      driver.close();
      driver.quit();
    }

    @Test
    public void searchGoogle() {
      Thread.sleep(1500);
      driver.get("http://www.google.com");
      // Find the text input element by its name
      WebElement element = driver.findElement(By.name("q"));
      // Enter search string
      element.sendKeys("Packt Mobile Automation");
      // Now submit the search
      element.submit();
      String searchHeader = driver.findElement(By.cssSelector
      ("H1")).getText().toLowerCase();
      Assert.assertTrue(searchHeader.contains("packt"));
    }
}
```

# Implementing a mobile test automation framework for emulators

Until now, we have learned how to set up mobile emulators and execute basic test cases on web browsers using Selenium. To create more robust test scripts that can be executed on hybrid and native mobile applications, we need to use other tools. Some commercially available tools that can be used to test with both emulators and simulators are **Experitest Seetest**, **Testplant Eggplant**, and **Ranorex**. However, these carry licensing costs and customers who generally go for implementing such tools do not use them against emulators and would rather opt for real devices or a cloud setup. So, with emulators, we can usually see prominent use of open source tools. In this chapter, we will cover a powerful automation support framework called **Appium**.

 We will cover **SeeTest** and other tools in the next chapter, which is about the automation of real devices. The same concepts can be used for simulators and emulators too.

Appium is built with a vision to support the testing of web applications, along with hybrid and native applications on Android and iOS devices using Selenium. So we can create scripts with Selenium Web Driver API and execute on both emulators and real devices. It supports Android, iOS, and also Firefox OS. So, the same scripts can be used for Android, FireFox OS and iOS applications and even for web applications and hybrid applications with minor adjustments like using **Remote Web Driver** instead of Web Driver. Appium primarily uses the `UIAutomator` library, which comes bundled with Android SDK to execute scripts on Android for Version 4.2 onwards. For earlier Android versions, it uses the `Selendriod` backbone. It also provides support to advanced features on mobile devices such as touch, multitouch, and pinch and zoom functions, which can be used for advanced test scenarios.

# The Appium architecture

Appium is an HTTP server written in **node.js** , which is used to create and handle multiple WebDriver sessions for different mobile platforms like iOS and Android.

Appium works in a similar way to Selenium server, in the sense that it has platform-specific handling for each incoming command from the main server to the client libraries.

The following figure demonstrates the architecture of Appium:

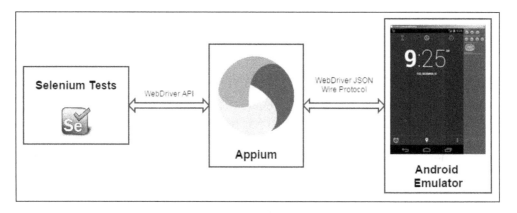

Appium accepts the commands from Selenium and uses **WebDriver JSON Wire Protocol** to translate them into an understandable format for UIAutomator.

# The Appium setup

In this section, we will learn about configuring Appium with mobile emulators and run scripts with the use of Selenium-based instructions on both hybrid and web applications. Since there are some OS-specific differences, we will discuss the setup on both Microsoft Windows OS and Apple Macintosh OS.

## Android on the Windows OS

To set up Appium, first of all, install the software with the following steps:

1. Download Appium from the `http://appium.io/`.

2. The latest version of Appium would be available for download on the page. Currently, the latest version of Appium is 1.2.4.1 (`https://bitbucket.org/appium/appium.app/downloads/AppiumForWindows-1.2.4.1.zip`).

The Microsoft .NET™ framework is a prerequisite to installing Appium on Windows OS, so make sure that it is installed on the machine prior to setting up Appium. If it is not installed during the Appium setup, it will be prompted for and you will need to install it before proceeding.

After installation is finished, click on the **Appium** icon inside the installation folder. It will open in a window similar to the one shown in the following screenshot:

Now, to configure Appium for a particular application and emulator, perform the following steps:

1.  Click on the **Android** icon to configure the various options. In this example, we have configured Appium for the ContactManager application with the absolute path of the .apk file provided under Application Path:

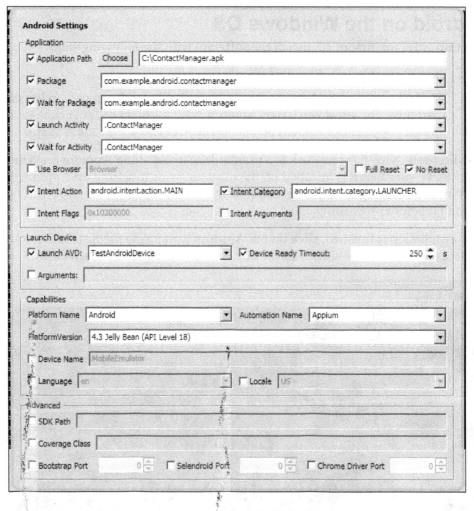

2.  If the AVD is already running, it will be displayed under the **Launch AVD** drop-down list. Select the appropriate one. It is important to ensure that the **PlatformVersion** is same as that of the Android AVD under test.

3.  For Android versions above 4.3, select **Appium** under the **Automation Name** drop-down list. For versions 2.3 and below, select **Selendriod**.

4.  In the **General Settings** window, enter **Server Address** as `127.0.0.1` and port as `4723` (default values). Check the following options: **Pre-Launch Application** and **Override Existing Session**. It is recommended to keep the **Check for updates** option unselected for faster execution:

**General Settings**

Server

| Server Address | 127.0.0.1 | Port | 4723 ⬍ |

☑ Check For Updates
☑ Pre-Launch Application
☑ Override Existing Session
☐ Use Remote Server
☐ Selenium Grid Configuration File

Logging

☐ Quiet Logging    ☐ Use Colors    ☐ Show Timestamps

☐ Log To File

☐ Log To WebHook

☐ Use Local Timezone

5.  To run automation scripts on emulators, the Android SDK or Apple XCode should also be installed and any necessary settings, like environment variables, should be set as explained in the previous steps.

6.  For automating a native or hybrid application, we also need to extract the `appPackage` and `appActivity` details. We can do this by either executing the `aapt` command to extract the metadata of an `.apk` file or by importing the `.apk` file in Appium, under the **Application Path** section. After importing the `.apk` file, the details would be autopopulated. To do this, execute the following command:

```
prompt>aapt -debug badging <qualified name of apk.apk>
```

Let's take a look at the following screenshot:

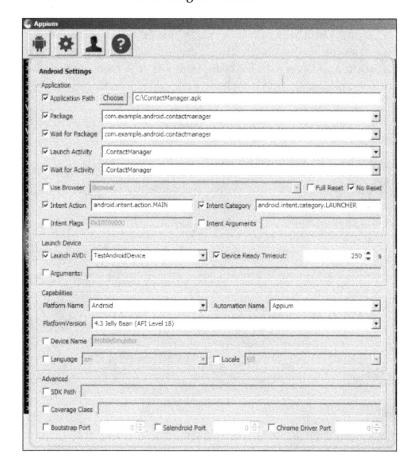

7.  After this, click on the **Launch** button at the top-right corner of the Appium window, which looks like the **Play** button. The console log will start loading in the window with details.

# Appium on the Mac OS

To set up Appium on Mac OS, first of all, install it by downloading the required .dmg file and performing the following steps:

1.  Start up the terminal.

2.  Enter the cd ~/ command to navigate to the home folder.

3.  Enter the touch .bash_profile command to create a new file.

4. Edit the `.bash_profile` file with any available editor (like TextEdit) by running the command `open -e .bash_profile`.

5. Add the following in the file: `export PATH=${PATH} ANDROID_HOME=<path_to_androidSDK_directory>/sdk/`.

6. Enter the `~/.bash_profile` command to reload the `.bash_profile` file and update any function to be added.

# Scripting and execution with Appium

In this section, we will understand the usage of Appium for automation with emulators and simulators. Let's first understand the object property extraction and scripting mechanism with Appium.

## Scripting and object property extraction

Appium comes inbuilt with the `Object Inspector` utility. After launching the AVD, click on the **Inspector** icon at the top-right corner of the Appium window. It will launch another window with the current active emulator that was launched by Appium. You can also send `Touch` commands and text input using this utility, but the most important usage is to provide the object hierarchy and other details required to create the object descriptions to be used in the script:

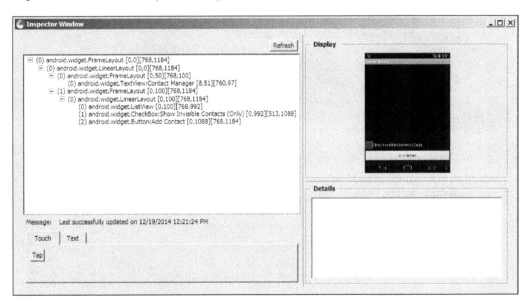

On navigating to another window, click on the **Refresh** button to capture the new screen objects and their hierarchies and properties.

# Execution

To execute scripts, we need to add a few additional capabilities to enable Selenium code to interact with the running emulators. Apart from inserting additional capabilities, we also need to trigger the remote web driver on the server URL and port that have been provided during the configuration of Appium. These can also be done programmatically, as shown in the following example.

# Code snippet

In the following example, we will automate an Android hybrid application called **ContactManager**. There are two major portions in this example. The first one is the `setCapabilities()` function, which takes the various capabilities as input variables and then uses them to set the capabilities that are then used by the second function, which uses **Android Remote WebDriver** to launch the application and perform the desired actions:

```java
import java.net.MalformedURLException;
import java.net.URL;
import java.util.List;
import org.openqa.selenium.By;
import org.openqa.selenium.WebElement;
import org.openqa.selenium.remote.DesiredCapabilities;
import org.openqa.selenium.remote.RemoteWebDriver;

public class AndroidAutomationWithAppium {
  public static DesiredCapabilities setCapabilities(String
  platformName, String platformVersion, String device, String
  deviceName, String avd, String app, String autoLaunch) {
    DesiredCapabilities cap= new DesiredCapabilities();
    cap.setCapability("platformName",platformName);
    cap.setCapability("platformVersion",platformVersion);
    cap.setCapability("device",device);
    cap.setCapability("deviceName",deviceName);
    cap.setCapability("avd",avd);
    cap.setCapability("app",app);
    cap.setCapability("autoLaunch",autoLaunch);
    return cap;
  }
```

These capabilities are used to set up the emulator at run time by calling the function with the required values, such as those used in the following code snippet:

```
public static void main (String args[]) throws
MalformedURLException {
  DesiredCapabilities cap1 = setCapabilities("Android","4.3",
  "Android","MobileEmulator","TestAndroidDevice",
  "C:\\ContactManager.apk","true");
  RemoteWebDriver remote = new RemoteWebDriver(new
  URL("http://127.0.0.1:4723/wd/hub"),cap1);
  remote.findElement(By.className("android.widget.Button"))
  .click();
  WebElement tableLayout = remote.findElement(By.className
  ("android.widget.TableLayout"));
  List<WebElement> tblRow = tableLayout.findElements(By.className
  ("android.widget.TableRow"));
  WebElement contactName=tblRow.get(3);
  WebElement contactPhone=tblRow.get(5);
  WebElement saveComponent=tblRow.get(8);
  contactName.findElement(By.className
  ("android.widget.EditText")).sendKeys("Feroz Louis");
  contactPhone.findElement(By.className
  ("android.widget.EditText")).sendKeys("0123456789");
  saveComponent.findElement(By.className
  ("android.widget.Button")).click();
}
```

# Capabilities of the Android emulator for versions lower than 4.2 and hybrid applications

The following code shows the capabilities of Appium for an Android emulator for versions lower than 4.2 (API 17) and hybrid applications:

```
{
  DesiredCapabilities cap= new DesiredCapabilities();
  cap.setCapability("platformName",platformName);
  cap.setCapability("platformVersion",platformVersion);
  cap.setCapability("device","Selendriod");
  cap.setCapability("deviceName",deviceName);
  cap.setCapability("automationName":"selendroid");
  cap.setCapability("avd",avd);
  cap.setCapability("app",app);
  cap.setCapability("autoLaunch",autoLaunch);
  return cap;
}
```

By using these capabilities, Appium will be launched with a Selendroid test session, instead of the default Appium test session. Note that since there are differences between the Selendriod API and Appium, there might be some changes required to scripts.

# Advantages of mobile emulators

The following are a few advantages of using emulators for mobile test automation:

- If we want to validate the original look, feel, and behavior of real devices without investing in one, then using an emulator is the best option.

- We can easily configure the hardware specification of the emulated real device that we want to use depending on the needs; for example, we can configure the operating system version, RAM, whether an SD card is present or not and its size, and many more for a lower-level validation with real-world conditions.

- Physical devices are cumbersome to maintain and manage, besides being costly, and any organization will have to keep its inventory updated with numerous releases of newer devices. With emulators, all these overheads can be eliminated in a cost-effective and secure way.

# Limitations of mobile emulators

As we have discussed briefly in *Chapter 1, Ensuring Five-star Rating in the MarketPlace*, since emulator programs cannot completely mimic a physical device, they cannot be relied upon for testing device hardware- or firmware-specific features like cameras, gyroscopes, accelerometers, network conditions, and geolocations. This is the most important limitation of mobile emulators. Also, with emulators, there are very limited tools that can be used to extract and create object identifiers. So, with the emulator-based automation approach, it is important to have experts of functional automation in the team, who can help create the object identifiers quickly.

Besides these, emulators face severe limitations with respect to user interface rendering and user interactions, for instance like swiping and scrolling. So, any test results obtained with emulators cannot be completely relied upon because emulators are also software that can be prone to bugs. When coupled with automation tools, there can be many discrepancies on an emulator that might not actually be an issue on a physical device. Hence, it is always advisable to do at least one round of physical device testing for any project. Also, all GUI defects must be verified on real devices prior to logging defects.

In essence, the use of emulators should be limited to the early stages of testing, such as unit testing, where it may not be economical to maintain a large repository of physical devices for all development team members.

# Troubleshooting and best practices

Working with emulators requires a lot of patience and practice. Making emulator programs work according to how you need them to perform, needs a solid programming mindset.

The following are some industry best practices and troubleshooting tips that will come in handy when working with emulators:

- Some platform emulators, like those for Blackberry and Windows, are not fully supported with any freeware automation tools, hence if it is a requirement to cover these platforms in the automation solution, then using emulators is not the right approach and it is best to go for other approaches.

- A best practice is to use a robust IDE, such as Eclipse, configured with both Android SDK and Selenium. For configuring Android on Eclipse, the ADT plugin should be utilized. Refer to this link for more information: `http://developer.android.com/sdk/installing/installing-adt.html`.

- Android SDK provides the **Android Debug Bridge** (**ADB**) and Android Emulator Console under the `platform-tools` folder, which gets installed along with the Android SDK. An important commands with ADB is `adb logcat`, which allows us to monitor all logs from the emulator program. Sometimes, this comes in handy when the Appium window doesn't provide sufficient details.

- Installing and uninstalling applications inside an emulator should be done programmatically, instead of manually. For this, `.bat` files can be created with the required batch commands (explained earlier in this chapter) and then these batch files can be used to create the emulator, launch it on the required machine, and then install the required mobile application. With this approach, we can also overcome the dependency to create a set of emulators manually on different machines. In such scenarios, it is also important to use options like `isAppInstalled`, `installApp`, and `removeApp`. The commands for Java are as follows:

```
driver.isAppInstalled("com.example.android.apk");
driver.removeApp("com.example.android.apk");
driver.installApp("pathtoapk/apkFileName.apk");
```

- The Android emulator allows us to create some hardware features like SD card size and RAM along with a spoofed GPS location, but with iOS simulator, this kind of hardware/firmware options are not allowed. Hence, if your scripts are to be executed on both Android and iOS, it is advisable to create emulators and simulators with default options only.

- Sometimes, while working with mobile applications that access network components, it is required to utilize the intranet network on the emulated device. This can be done by using the `avd` command, as follows:

```
prompt> emulator -avd <AndroidEmualtorName> -http-proxy <http://
proxy:8080>
```

- To extract the software version and bundle ID key values from the iOS `.ipa` file that are used while configuring Appium, perform the following steps:

  ○ Right-click on the application file and select the `show package contents` option

  ○ Open the `plist` file from the application contents with any text editor to get the bundle ID

  ○ Alternatively, you can open its `iTunesMetaData.plist` file in an editor to extract these values

- To automate, one of the key components required are application object properties and values. Even though Appium provides the object inspector, it is not very reliable, especially on Windows OS. Let's take a look in more details:

  ○ For mobile web applications, it is advisable to use the user agent approach to extract the application object property values like ID, XPath, CSS selectors, and others that can then be used in the selenium scripts.

  ○ For native or hybrid applications, apart from the default Appium object inspector, additional tools that are commercially marketed can be used, like **droidInspector**.With Selendriod, one can use **Selendroid Inspector** for earlier versions of Android. After you have extracted the values, you can use the same on iOS too with minor modifications. However, in most cases, you would be required to create XPath manually in a trial-and-error manner. Another handy tool is the Android **UI Automator Viewer**, which comes bundled with the Android SDK. It can be accessed from the `tools` folder, under the Android SDK folder.

# Summary

In this chapter, we learned about the usage of emulators and simulators for automation. We saw how to set up mobile emulators, which are often thought of as the trickiest part in mobile automation. We then learned about the most commonly used automation tools and support frameworks. The code snippets in this chapter also came in handy while we learned how to work with real devices.

With the implementation of an automation framework like Appium, we have been able to learn the nuances of mobile automation with emulators. But, at the same time, we need to be aware of the limitations of this technique and this technique should not be relied upon for complete verification and validation of a mobile application. The only reliable way to validate mobile applications is to use real devices.

In the next chapter, we will focus on this aspect and will learn about the automation techniques that come prepackaged in real devices.

# Automating Physical Devices

**5**

In this chapter, you will understand the techniques required to automate a mobile application test with physically present real mobile devices. This approach is one of the most reliable ways to validate any mobile application. It allows for maximum flexibility with respect to testing all types of scenarios, as it is this approach that provides maximum control of the devices out of all the others. Apart from the brief introduction about this technique, in continuation to what we learnt in *Chapter 1, Ensuring Five-star Rating in the Market Place*, in this chapter you will understand the various tools that can be used for automation with real devices and also learn the most effective ways to automate with this technique. While discussing the implementation, we will also learn the real world practical limitations of this technique and some ways in which you can overcome these shortcomings.

We will cover the following topics:

- Process to set up and enable real devices for mobile automation
- Code snippets with working examples
- Advantages and limitations of this technique
- Troubleshooting and best practices

# Getting started

Use of physically present real devices for test automation is the most obvious choice for most organizations, given that this is also one of the most common ways of manually testing of applications. This technique is especially used during the early stages of testing a lifecycle, when developers need an easy and quick way to validate their application on devices after making changes. As most organizations begin by replicating their manual testing process in automation, it is also quite often the case that teams start automating on them. There are only a few open source and freeware automation tools that provide the facility to do automation on real devices. Hence while selecting this approach, it is very important for project managers to do a thorough analysis of the type of test cases that need to be automated. In most cases, only functional automation can be achieved with open source tools such as **Appium** and **Selendroid**. So, if there are any test cases that need interruption scenarios, layout verifications, or any other features that need advanced user interactions such as multi-touch inputs, then it is better to invest in some **Commercial Off The Shelf (COTS)** tools.

The most prominent tools in the COTS category that provide real device automation capability are:

- ExperiTest (SeeTest)
- TestPlant eggPlant Mobile /eggOn
- Jamo Solutions M-eux Test
- Zapfix
- SOASTA
- Ranorex

In this chapter, we will learn about automation on real devices with Appium and SeeTest, which are one from each of the open source and COTS categories of tools respectively. The principles illustrated with these tools are generic and will allow us to develop our understanding of the automation process with real devices, which can be easily applied to other tools listed in the preceding section.

# Automation technologies with real devices

It is very important to have a clear understanding of the automation technology that a tool uses in order to be able to select the right tool, as per a project's requirements. The importance of this stems from the fact that many a time, it might not be feasible to get the source code of an application to enable automation with tools that need instrumentation of code; in other cases, it might not be possible to test the application with a Wi-Fi connection while doing tethering simultaneously. So here, we will learn about the various automation technologies and their implications on any project requirements.

Basically, there are the following technologies:

- Optical recognition
- AUT code instrumentation
- Accessing native device methods

# Automation with the optical recognition technology

In this technology, the tools use an image comparison and mapping algorithm to capture the application objects during recording, and then identify the correct object from the displayed application controls. The automation scripts are sometimes based on screen coordinates of elements. For example, tapping a button on the screen is achieved by tapping the coordinates such as $x=45$, $y=55$. As this technique of object identification basically relies on object extraction as per application bitmap images, it is very brittle and needs recreation of scripts for different devices, which vary with their screen resolution, screen size, and even the orientation changes.

So given this limitation, in some tools, object recognition is further augmented with the **Optical Character Recognition (OCR)** mode, which allows for extracting text out of the images displayed on the application screen from the device and then doing a text-based identification. Tools such as **EggPlant** and Selenium with **Sikuli** use this kind of object recognition. A major limitation of this technology is that it may need jail-breaking or rooting of device operating systems. This may not be desirable for organizations that have a strict security policy with respect to **Mobile Device Management (MDM)**.

# Automation with AUT code instrumentation

Many tools perform object recognition by enabling automation of controls within the AUT by recompiling automation-tool-specific libraries along with the application code. This is especially important for Android devices where fragmentation is rampant and there are numerous screen size variants available in the market.

So this technology is not suitable for projects where testing teams are not allowed access to application code or with organizations that may not be willing to rebuild their applications with external or third-party libraries. Tools such as **Jamo solutions** and **Monkey talk** have this kind of object recognition feature without the need for rooting or jail-breaking. **Zapfix** allows users to either do rooting of devices or to instrument the application code. **Ranorex** requires code instrumentation also, which it then utilizes with its **RanoreXpath** object recognition mechanism, which is based on both the object properties and the object hierarchies along with relative screen coordinates.

# Automating with native device methods

Some automation tools can access native device methods to replicate end user interactions. In many cases, this may require jail-breaking or rooting of device operating systems. But typically with commercial tools, this is not a prerequisite. Here, object properties such as the tags and IDs are used to identify objects and replay the end user interaction with the device OS and shell commands. The device OS kernel still has still restricted access thereby ensuring security. Most COTS tools use this technique now because this technology is the least pervasive to application code and is also in-line with enterprise security concerns of large organizations. So leading tools such as **SeeTest** and **SOASTA TouchTest** from the COTS category and Appium from the open source category are based on this object recognition and replay technology.

# Object Identification with leading tools

In the following table, object recognition techniques used with various tools are listed for quick reference:

| Tool | Object Recognition Mechanism | Application Code Instrumentation Requirement | Jail-Breaking or Device Rooting Requirement |
|---|---|---|---|
| SeeTest | It identifies objects based on Native ID and DOM. It also provides support for image-based recognition. | Yes (It can work with non-instrumented apps to a limited extent) | No |
| EggPlant | This is Image and OCR-based object recognition. | No | No |

| Tool | Object Recognition Mechanism | Application Code Instrumentation Requirement | Jail-Breaking or Device Rooting Requirement |
|------|------------------------------|-----------------------------------------------|----------------------------------------------|
| Jamo M-eux | This is Native ID-based. | No | No |
| Zapfix | This is an object ID based on instrumented code. | Yes | Yes (if instrumentation not done) |
| SOASTA | Native ID based | No | No |
| Ranorex | This is based on relative screen coordinates and object hierarchies along with native ID. | Yes | No |
| Appium | This tool is for native object interactions based on UIAutomation for Android and System Trace with XCode on iOS. | No | No |
| Selenium | This works through code injection in browsers. | No | Yes |
| Sikuli | This is image-based recognition. | No | No |
| Monkey Talk | Image. | Yes | No |

The mobile automation tool list is ever expanding, and this table is not exhaustive by any means. It is only an indicative list to help determine the differences between leading mobile automation tools' object recognition techniques. The tool selection should not be based solely on this criterion.

# Automation on real devices with various tools

In this section, we will understand how to automate using various sets of tools such as Appium from the open source category and SeeTest from the COTS category. The reason for using SeeTest is that it is one of the most versatile mobile automation tools. It not only supports all major mobile platforms but also provides support for all object recognition technologies. So with this tool, we will be able to cover all object recognition technologies and the principles learned, so they can be applied to individual tools that work with any one of these object recognition technologies.

# Automation with Appium on real devices

As we learnt in the previous chapter, Appium is an open source tool that supports both Android and iOS automation, and also real devices along with emulators and simulators. It is the leading open source tool when it comes to automation with real devices.

To enable automation with Appium on real devices, we need to enable USB debugging, and the respective device driver needs to be installed on the machine from which the tool is going to be executed. For Android, device drivers can be found from the respective manufacturer pages or from `http://developer.android.com/tools/extras/oem-usb.html`.

For iOS, the iTunes software needs to be installed on the machine which is required to be used as the connection controller.

- Every device has a serial number which is used to identify it uniquely on the system. To retrieve the serial ID of the device connected to the system, run the following command:

  ```
  prompt>adb devices
  ```

- Download the Android server from the following location and save it in the `platform-tools` directory under the `android-sdk` directory:

  ```
  http://selenium-release.storage.googleapis.com/index.html
  ```

- Either install the Android server manually on the device or run the following command:

  ```
  prompt>adb -s <serialId> -e install -r android-server.apk
  ```

- Start the Android **WebDriver** application, by running the following command:

  ```
  prompt>adb -s <serialId> shell am start -a android.intent.action.MAIN -n org.openqa.selenium.android.app/.MainActivity
  ```

- Connect your Android device with the PC and make sure the device's **Developer** option is **ON** and also the **USB debugging mode** option is checked.

- Install the `adb` driver which you have downloaded before.

- Now start the Android server using the `adb start-server` command from the command prompt.

- Check for your device with the following command:

```
prompt> adb devices
```

All connected devices' serial IDs will be displayed.

# Code snippet for Appium with real devices

In the following example, we will automate an Android hybrid application called **ContactManager**. There are two major functions in this example. The first one is the `setCapabilities()` function which takes the various capabilities as input variables, and uses them to set the capabilities that are then utilized by the second function, which uses Android Remote WebDriver to launch the application and perform the desired actions on real devices:

```java
import java.net.MalformedURLException;
import java.net.URL;
import java.util.List;
import org.openqa.selenium.By;
import org.openqa.selenium.WebElement;
import org.openqa.selenium.remote.DesiredCapabilities;
import org.openqa.selenium.remote.RemoteWebDriver;

public class AndroidAutomationWithAppium {
  public static DesiredCapabilities setCapabilities(String
  platformName, String platformVersion, String device, String
  deviceName, String avd, String app, String autoLaunch) {
    DesiredCapabilities cap= new DesiredCapabilities();
    cap.setCapability("platformName",platformName);
    cap.setCapability("platformVersion",platformVersion);
    cap.setCapability("device",device);
    cap.setCapability("deviceName",deviceName);
    cap.setCapability("avd",avd);
    cap.setCapability("app",app);
    cap.setCapability("autoLaunch",autoLaunch);
    return cap;
  }
```

Now we can use the `setCapabilites()` function from the `AndroidAutomationWithAppium` class in the main function to launch a device with the required application and perform further functions. Being parameterized, this function ensures that we can reuse it with the only changes in data as per the device and application details. We can further link it to an external input file using Excel or Database and then it will become a completely hybrid framework, as shown in the following code:

```
public static void main (String args[]) throws
MalformedURLException {
  DesiredCapabilities cap1 =
  setCapabilities("Android","4.3","Android","Device",
  "TestAndroidDevice","C:\\ContactManager.apk","true");
RemoteWebDriver remote = new RemoteWebDriver(new
  URL("http://127.0.0.1:4723/wd/hub"),cap1);
  remote.findElement(By.className("android.widget.Button"))
  .click();
  WebElement tableLayout = remote.findElement(By.className
  ("android.widget.TableLayout"));
  List<WebElement> tblRow = tableLayout.findElements
  (By.className ("android.widget.TableRow"));
  WebElement contactName=tblRow.get(3);
  WebElement contactPhone=tblRow.get(5);
  WebElement saveComponent=tblRow.get(8);
  contactName.findElement(By.className
  ("android.widget.EditText")).sendKeys("Feroz Louis");
  contactPhone.findElement(By.className
  ("android.widget.EditText")).sendKeys("0123456789");
  saveComponent.findElement(By.className
  ("android.widget.Button")).click();
  }
}
```

# Capabilities for Android devices with versions below 4.2

The following code shows the capabilities of the Appium for an Android emulator for versions less than 4.2 (API 17) and Hybrid applications:

```
{
  DesiredCapabilities cap= new DesiredCapabilities();
  cap.setCapability("platformName",platformName);
  cap.setCapability("platformVersion",platformVersion);
  cap.setCapability("device","Selendriod");
```

```
cap.setCapability("deviceName",deviceName);
cap.setCapability("automationName":"selendroid");
cap.setCapability("app",app);
cap.setCapability("autoLaunch",autoLaunch);
return cap;
}
```

# Automation with SeeTest on real devices

It is important to note that many mobile automation tools provide add-ons for leading automation tools like UFT, which enable automation of mobile applications almost in the same way as that of web-and desktop-based applications. We will learn about this approach with cloud-based solutions as it is almost the same as that with real devices minus, the changes required for cloud access and control.

In this section, we will learn how to use a leading COTS tool **ExperiTest SeeTest**. The principles that we learn in this chapter are valid for use with emulators as well. SeeTest allows automation with Android, iOS, Windows, and even Blackberry and Symbian devices.

## Configuring a SeeTest license

After completing the installation, add the license file that you have been provided (if it's not provided, use a trial license) by navigating to **Help | License Information** and then adding the relevant license type. The license information window is shown in the following screenshot:

**Seat license activation** works on only one machine for which the installation has been done, and **Floating license activation** can be used like the typical UFT floating license, with which only a predefined number of users can use the tool at any given point of time, but there is no restriction on the total number of tool installations on any given network. So floating license type is especially important for teams that work in a geographically distributed way.

It is important to note that configuring a floating license would require setting up a server that would act as the central node that verifies the licenses in use at any given point of time within the network.

# SeeTest tool configuration

Before we learn to configure SeeTest, let us understand the important components of the tool.

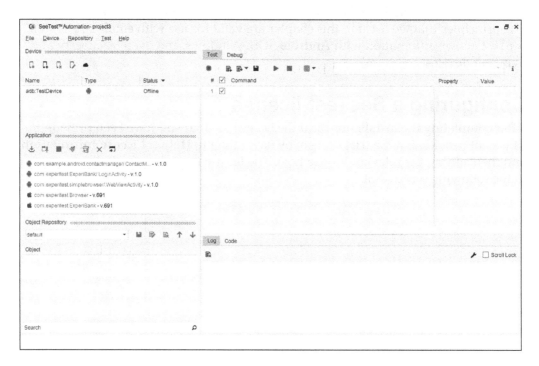

The main sections in the tool window are:

- **The device section**: It allows us to control and modify device and record mode settings.
- **The test section**: It shows the test script that is recorded on the device and also allows us to execute it for testing.

- **The application section**: It shows various instrumented applications available for testing and also applications that are available on connected devices which can be used for testing.

- **The object repository section**: It shows the objects being recorded on the applications and their property-value pairs in the right-hand side section of the test window.

- **The log console**: It shows a runtime log of all activities on SeeTest.

- **The code tab**: It provides the ability to export the script in any code of choice:

  ° Java (junit3)
  ° Java (junit4)
  ° Java TestNG
  ° Java WebDriver (Selenium)
  ° C# NUnit
  ° C# MSTest
  ° UFT (QTP) VBScript
  ° RFT (Java)
  ° VB.NET
  ° TestComplete VBScript
  ° Python
  ° Perl
  ° Ruby

Now we will learn about the step-wise approach to SeeTest automation after installation:

1. Configure real mobile device for automation.
2. Configure application (instrumentation).
3. Create scripts.
4. Export code and integrate with execution environment of choice.

# Configuring a real mobile device for automation

To configure devices for automation, we need to follow different techniques for different operating systems. These techniques are universally the same regardless of the automation tool.

## Android devices

To enable an Android mobile device for automation, first of all we need to download and install the Android USB drivers. Then we need to enable the **USB debugging** option in the devices. For Android devices with a version number less than 4.0, we need to go to **Settings** | **Applications** | **Development** and set the **USB debugging** checkbox to enabled. Also enable the **Stay awake** option so that the device doesn't get frequently locked during execution.

For Android devices with version 4.0 and above, select the **OK** option when the **Allow Device Option** popup appears after connecting the device to the user machine:

# iOS-based devices

There are two modes to enable automation with iOS devices. In one mode, we can also control the device spring board apart from controlling the device with SeeTest.

In both modes, iTunes needs to be installed, which automatically updates the USB drivers.

For the first mode, after connecting the iOS device, the **Enable non-instrumented mode** checkbox needs to be selected from the **Connect Device** option in SeeTest. Then, after connecting the device, a **Device Enablement File (DEF)** needs to be generated. This can either be generated from a third-party service like `denabled.com` or by creating the iOS developer profile with **XCode**. Then this DEF file needs to be imported in SeeTest to enable device automation. In the second mode, we can proceed simply by keeping the **Enable non-instrumented mode** checkbox disabled while connecting the device.

# Windows-based devices

For Windows phones, the minimum OS requirement for a user machine is Windows 8.

Follow these steps to configure a Windows device:

1. Install Windows Phone SDK on the user machine from `http://dev.windows.com/en-us/develop/download-phone-sdk`.

2. Register the device under test for development, which requires a developer account. This needs to be done from `http://msdn.microsoft.com/en-us/library/windows/apps/ff769508(v=vs.105).aspx`.

3. Copy the application `.xap` file on the device to the user machine, which will be used as a communication bridge between SeeTest and the application under test on the Windows phone device, and instrument it.

4. After this, the device can be connected via a USB cable, and it will be detected automatically by SeeTest.

# Blackberry devices

SeeTest is one of the few automation tools that allows automation with Blackberry devices. However, Blackberry devices need a different kind of setup for enabling automation. Follow this stepwise approach for Blackberry phones:

1.  Install the Blackberry desktop software for the PC from `http://us.blackberry.com/apps-software/desktop/`.

2.  Go to the **Device** tab in the SeeTest window and select the Blackberry phone from the options within **Add Device**.

3.  For the Blackberry device, SeeTest will show a pop-up window (for the first time only) to deploy the agent. In this window, enter a suitable password and click on the **Deploy Agent** button:

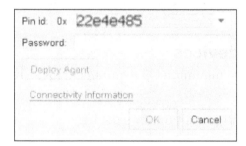

4.  After this, on the Blackberry device, click the **Blackberry Remote Agent** application icon that is deployed and approve any special permission as necessary.

5. After completing this process, an ExperiTest Blackberry remote agent screen will be shown on the device, such as in the following screenshot:

6. After configuring the agent, we need to configure additional settings on the device by going to **Options | Device | Application Management | BB Device Agent | Edit Permissions** and selecting **Allow** for all the options shown on this screen:

7. Now go to the **ExperiTest BlackBerry Remote Agent** application again and launch it.

8. As a final step, choose Blackberry from the device list shown in the **Open Device** option, which will launch the configured device.

# Instrumentation of a mobile application

Instrumentation is the process which enables automation tools to access application controls and perform interactions programmatically. In this section, we will learn to instrument our **Application Under Test** (**AUT**) for enabling automation with SeeTest.

Different tools come with tool-specific ways to instrument the application, but OS-specific steps remain the same. So this section will provide a good overview to understand the prerequisites.

## Android platform

Follow these steps to instrument an Android Native or Hybrid application:

1. Ensure that the prerequisite JDK 6 or higher is installed on the user machine.

2. Ensure that the application `manifest.xml` file has the required permission entry to enable Internet access, as shown here:

```
<uses-permission  android:name=""android.permission.INTERNET""/>
```

3. If not, then add this entry in the XML file.

4. From the SeeTest window, under the **Application** section, use the **Import/Sign Application** option to import the application's `.apk` file.

5. Select the desired option to import the .apk file (either from the local disk or directly from the device) and click the **OK** button.

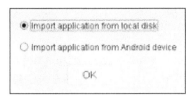

6. This option will open a dialog box with the listed application, depending on the option selected, and will allow to import it. As during the import process, the application is instrumented with SeeTest-specific code, it might take around five minutes to complete the process.

7. After the instrumentation process is completed, SeeTest will prompt the user to install the application on the target device.

8. If not done immediately, the application installation can also be done from the device reflection (the program window which displays the device screen with additional controls) that allows a user to perform actions directly on the reflection rather than having to perform the steps manually on the connected device) by using the **Install Application** icon.

After successful installation, the **Launch Application** icon can be used anytime to launch the instrumented application on the target device.

## iOS platform

The iOS ecosystem demands that unsigned applications are not allowed to be installed on devices, unless jail-broken; hence the application needs to be re-signed duly after the instrumentation process, because it makes some changes to the application package. For re-signing the application, you need to provide a p12 certificate to sign the application along with a debug provisioning profile matching the provided certificate. These files are available from a machine with duly configured XCode to sign applications. Usually the development team of an application can be approached for this one time activity.

Follow the same steps as those of Android to instrument an iOS Native or Hybrid application, and after instrumentation, SeeTest will prompt for an iOS instrumentation configuration with a dialog box as shown here:

Provide the **Signing certificate** and matching **Provision profile** files along with the certificate password to complete the process. After completion, the instrumented application can be installed on any connected iOS device.

## Windows platform

For Windows applications, the application XAP file needs to be imported. The remaining steps are the same as those mentioned in the Android section.

To enable identification of custom UI elements in Windows applications, the required custom control should be added in the Configuration XML file as a `Control` element in the following structure:

```
<Control class="CS.Windows.Controls.WatermarkTextBox">
<Click methodName="Focus" />
<Value />
</Control>
```

Also, the C# full class name that is used to implement the custom control should be provided as shown here:

```
class="CS.Windows.Controls.WatermarkTextBox"
```

For example, if a custom UI element is required to support the `Click` command in SeeTest, then the `Click` element should specify the method name implemented in the application code that will be called when the click action is performed on this element:

```
<Click methodName="Focus" />
```

If the custom UI element should support `ElementSetProperty` and `ElementGetProperty` commands, the `Value` XML element should be added:

```
<Value />
```

After the custom controls are defined in the `AutomationConfig.xml` file, perform the following steps, and save the application, and reimport to instrument the application.

## Blackberry platform

For Blackberry, SeeTest only supports a manual build process to instrument the application under test. For this, first the `Experitest BB Lib` file needs to be added as an external JAR file to the Blackberry application project. The JAR file download location is `https://experitest.s3.amazonaws.com/ExperitestBBLib.jar`.

After adding this as an external JAR file, two lines of code need to be added to the application code in the main Java function and the `onClose()` function:

```
public static void main(String[] args) {
  ExperitestLibManager.start();
  //the ExperitestLibManager will now always start along with the
  app
  TestDemoApp myApp = new TestDemoApp();
  myApp.enterEventDispatcher();
}
public boolean onClose() {
  ExperitestLibManager.quit();
  //Closing the ExperitestLibManager thread
}
```

In case of any nonstandard UI elements, the extension can be added as shown in the following code snippet:

```
public static void main(String[] args) {
  ExperitestLibManager.start();
  //the ExperitestLibManager will now always start along with the
  app
  ExperitestLibManager.addExtension(new Extension() {
    public boolean isAccepted(Object myObj) {
      return (myObj instanceof ListStyleButton);
    }
    public String getText(Object myObj) {
      return ((ListStyleButton) myObj.getLabel();
    }
  });
  TestDemoApp myApp = new TestDemoApp();
  myApp.enterEventDispatcher();
}
```

The steps mentioned till now are only one time setup activities to enable automation. Ideally, when required by the test scenario, the **Install** and **Launch** options should be performed programmatically by using the `install` and `launch` commands provided by the tool.

## Mobile web applications

For mobile web applications, the default browser applications such as Safari on iOS may be instrumented as an AUT, and then the desired mobile web application may be launched. Ideally, the following launch command should be added to the script at the beginning:

```
Launch(activity, instrument, stopIfRunning)
```

In the `Activity/URL` property of this command, set the value `safari:http://{URL of the website to be displayed)`, for example: `safari:http://www.packt.com`.

Apart from supporting the default browsers, SeeTest also provides a pre-instrumented browser application which can be pushed to the **Device Under Test (DUT)**, and automation of mobile web applications can be carried out with the Web Recognition method. Since this browser application is specifically designed to enable SeeTest automation, it is more robust when it comes to identifying Web Dom properties.

This browser can be installed from the following locations for the respective OS:

- For iOS, the location is `\bin\ipas`, and the file name is `Browser.ipa`
- For Android, the location is `\bin\adb`, and the file name is `simplebrowser.apk`
- For Windows, the location is `\bin\adb`, and the file name is `simplebrowser.xap`

## Creating scripts

To create scripts in SeeTest, first of all a basic scenario flow needs to be recorded in the tool, which serves the dual purpose of providing a skeleton of the scripts as well as capturing the required objects in the object repository. After recording of the basic flow scripts, they need to be edited to add validations, and you need to provide additional object properties for robust object recognition.

After the basic flow is captured and necessary modifications are made to the scripts, they need to be exported in the execution framework of choice so that logical steps can be added.

As we learned earlier, SeeTest provides support for all types of object recognition modes:

- Dynamic Recording
- Native Recording
- Web Recording
- Image and Text (OCR) Recording

Before we learn about these object recognition modes in more detail, let us first learn how to create scripts with SeeTest.

First of all, create a new project in SeeTest so that the files can be saved and referred to later, using the **New Project** option from the **File** menu of SeeTest.

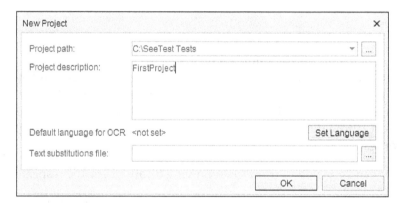

During the process of project setup, language setting can be done for OCR object recognition mode, and any text substitutions file can also be provided, which may be AUT-specific. These files can generally be obtained from the application **Software Requirement Specification (SRS)** document.

After connecting the DUT with the user machine, having SeeTest installed and activated, and the AUT properly instrumented, press the red button in the SeeTest window to activate the recording mode.

This will open a dialog box to configure the device on which recording needs to be activated, and the application for which recording needs to be done, along with the recording mode configuration.

It is generally recommended to keep the **Dynamic Recording** checkbox selected, which activates object recognition in Dynamic mode. After pressing the **Start** button, a **Preparing For Recording** dialog box with a status bar would be displayed. When the status bar has fully loaded, it means that SeeTest has successfully been able to set up a connection with the AUT on the DUT. After this stage, the application would be launched automatically on the reflection window of the DUT on the user machine and recording would be activated. It is important to note that all steps need to be performed only on the reflection window and not on the device. For emulators, there is no reflection launched and steps may be directly performed on the emulator window.

To add validation points with SeeTest during recording, right click on the required object.

In this example, we have used the **SimpleBrowser** application to launch Google and navigate to the Packt Publishing website. Please note that during the recording phase, no steps are generated at runtime as with other tools like UFT. SeeTest analyzes all the recorded steps we click on the **Stop Recording** button. A progress bar is displayed, after which SeeTest generates the recorded steps and captured objects.

The steps are shown in the **Test first** section, and the objects are shown in the **Object Repository** section.

Before making any changes to the scripts or exporting the code for integration with the external execution environment, it is generally recommended to do a dry run to check whether object recognition and step replay is performed properly on the device. For this, enable the **Launch Report** option of SeeTest from the **Help** menu and click on the **Play** button. After execution, a HTML report will be launched in the browser window with screenshots of the device during execution.

Now let us learn how to configure various recording modes with SeeTest in more detail:

- **Dynamic Recognition**: This mode is supported only for instrumented applications. During recording it captures both image and native properties of the objects. To enable this mode, select the options at the start of the recording session as shown in the following screenshot:

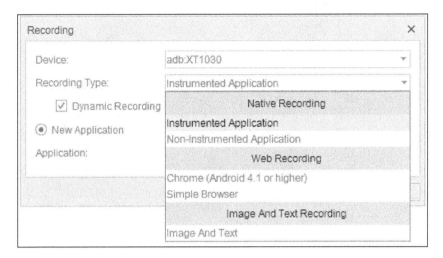

- **Native Recognition**: To enable this mode, select the **Native Recording** options in the recording window. Alternatively, after the recording is completed, this mode can be activated while adding new objects to the object repository using the **Object Spy** option:

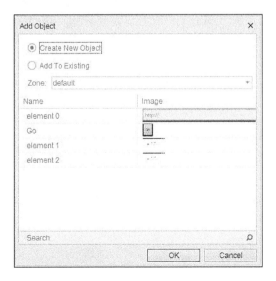

- This option can also be enabled for pre-recorded objects by selecting the checkbox for **Native** properties:

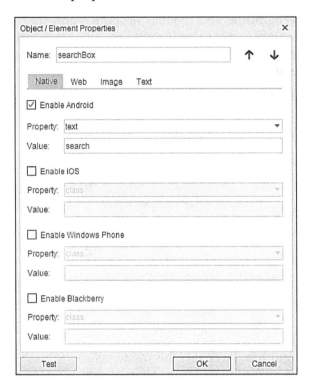

- **Web Recognition**: In this mode, objects are identified based on the Web Dom properties. It is not limited to web applications only, and this recognition mode can work for hybrid applications as well, which utilize HTML 5 contents. In this mode, the entire object hierarchy is used to identify the objects, and it can also work with XPath identification mechanism.

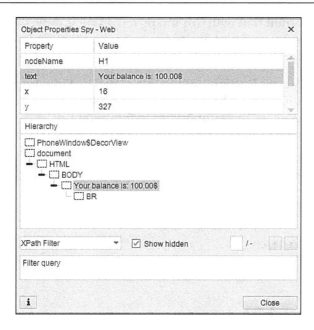

So this mode is most compatible when used with WebDriver.

- **Image Recognition**: As the name suggests, this recognition mode is based on matching images as displayed in the application:

So it is important that with this mode, recording is done by clicking at the center of images, and also at a normal speed, so that optimum information may be extracted while saving images.

- **Optical Character Recognition**: In this mode, text is extracted from the images displayed. The language can be selected while configuring the project, and SeeTest supports a lot of different character sets. While configuring the text extraction, color filter and sensitivity needs to be provided until the text is properly extracted and shown in the following box:

We can enable multiple modes of object recognition for the same object, and SeeTest will identify the object based on any of the techniques.

# Exporting scripts for integration with other execution environments

As explained in the previous sections, SeeTest provides an inbuilt facility to export code for the scripts recorded to a set of execution environments. In this section, we will learn about the process of exporting code and integrating it with external execution tools. For this explanation, we will use examples of both Selenium and UFT with the Packt Publishing script we created in the previous section.

To export code after the creation of script, use the **Code** section tab beside the **Log** section after selecting the desired execution environment.

# WebDriver code snippet

In the following code snippet, we will learn about the use of Selenium WebDriver with SeeTest. This code is shown as exported from Selenium with minor modifications:

```
import com.experitest.selenium.*;
import org.openqa.selenium.By;
import org.junit.*;
public class first {
  private String host = "localhost";
  private int port = 8889;
  private String projectBaseDirectory = "C:\\SeeTest Tests";
  protected MobileWebDriver driver = null;
  @Before
  public void setUp() {
    driver = new MobileWebDriver( host, port,
    projectBaseDirectory, "xml", "reports", "first");
  }
  @Test
  public void testfirst() {
    driver.setDevice("5554:TestAndroidDevice");
    driver.device().sendText("www.google.com");
    driver.device().sendText("{ENTER}");
    driver.findElement(new ByObject("default", "element
    6")).click();
    driver.findElement(new ByObject("default", "element
    7")).click();
    driver.device().sendText("packt publishing");
    if(driver.findElement(new ByObject("default", "element
    8")).waitFor(30000)) {
    }
    driver.findElement(new ByObject("default", "element
    8")).click();
    if(driver.findElement(new ByObject("default", "Packt
    Publishing")).waitFor(30000)) {
    }
    driver.findElement(new ByObject("default", "element
    10")).click();
  }
  @After
  public void tearDown() {
    driver.generateReport();
  }
}
```

This code can be used directly with Selenium, if no existing framework is in place, but if there is a pre-existing framework being used, then minor modifications, such as parameterization of test data, and linking to external Excel files, and reporting statements that are specific to the automation framework, may be required.

## UFT code snippet

Here is the code snippet for UFT. Note that it requires setting up an instance of **DotNetFactory** to integrate SeeTest with UFT. UFT in this case is used purely as an execution controlling mechanism, and all the object recognition information resides within SeeTest only:

```
Set client = DotNetFactory.CreateInstance("experitestClient
.Client", "C:\\Program Files\\Experitest\\SeeTest\\clients\\
C#\\imageClient.dll")
client.Connect "127.0.0.1", 8889
client.SetProjectBaseDirectory "C:\\SeeTest Tests"
client.SetDevice "5554:TestAndroidDevice"
Report
client.SendText "www.google.com"
client.SendText "{ENTER}"
client.Click "default", "element 6", 0, 1
Report
client.Click "default", "element 7", 0, 1
Report
client.SendText "packt publishing"
Report
If StrComp(client.WaitForElement ("default", "element 8", 0, 30000
), "True") = 0 Then
  Reporter.ReportEvent, micPass, "Element Existence", "Element
  Exists"
Else
  Reporter.ReportEvent, micFail, "Element Existence", "Element
  Doesn't Exists"
End If
client.Click "default", "element 8", 0, 1
Report
If StrComp(client.WaitForElement ("default","Packt
Publishing",0,30000),"True") = 0 Then
  Reporter.ReportEvent, micPass, "Element Existence", "Element
  Exists"
Else
  Reporter.ReportEvent, micFail, "Element Existence", "Element
  Doesn't Exists"
```

```
End If
client.Click "default", "element 10", 0, 1
Report
```

The `Report` function is used to send reporting information to the SeeTest log window, and if required it can be removed and only the in-built UFT reporter functions may be utilized, as shown in the following code snippet:

```
Sub Report()
Dim logLine, outFile, status, errorMessage
logLine = client.GetResultValue ("logLine")
outFile = client.GetResultValue("outFile")
status = client.GetResultValue("status")
If StrComp (status, "True") = 0 then
  Reporter.ReportEvent micPass, logLine, "", outFile
Else
  errorMessage = client.GetResultValue("errorMessage")
  Reporter.ReportEvent micFail, logLine, errorMessage, outFile
End If
End Sub
```

# Advantages and limitations of this technique

This mobile automation technique is closest to a real-world scenario as it allows for testing on real devices as well as with real networks such as GSM, CDMA, or Wi-Fi. As the devices are physically present for automation, they allow for their maximum control and quick and easy access to service devices such as upgrading the software and operating systems. Because this methodology is free of any other intervening tools that are needed for remotely controlling devices, or emulators and simulators that merely replicate the behavior of devices, it is the most dependable of all automation techniques. However, it has a few disadvantages as well, such as:

- Cost can be high due to multiple devices needing to be procured for different teams and testers, especially in the case of teams being spread across geographically distributed locations, thus limiting the sharing of devices.

- Maintenance of real devices and ensuring physical security of devices is an overhead.

- With this technique, script maintenance can be delayed if testing cycles are overlapping between functional and automation teams and both teams need to utilize the same set of devices for testing.

- In some cases, devices may need to be rooted or jail-broken to enable automation, and thus the automation results may not be entirely reliable.

- This technique is the most dependent on manual intervention as physical devices may need to be shared across teams for different purposes at different times, and this could mean that one time setup activities would need to be performed repeatedly.

- Many tools that support real device automation may require code tooling or instrumentation, which essentially alters the application code to some extent. The results thus obtained may need to be individually verified on non-instrumented versions before reporting any bugs, and this would mean that additional effort needs to be invested with this technique.

- One major limitation of this technique is that testing can only be carried out on devices and operating systems that are commercially available in the market, and hence beta testing is not feasible.

- Many real world scenarios such as battery state changes, incoming or outgoing calls or SMS, device orientation, and gyroscope and geographic location controls are not allowed to be mocked up and thus are very difficult to control programmatically. Hence, although with this technique we utilize real devices, many real-world scenarios are tedious to automate.

# Troubleshooting and best practices

Let's take a look at the following troubleshooting techniques and best practices:

- With real device automation, we need to set up connections via USB, Wi-Fi, or Bluetooth. Out of all these methods of connection, USB is the most effective and has the least time delay with respect to script execution in addition to being the most reliable one because it has no intervention with any other network.

- Real devices need to be configured for the least amount of manual intervention. For this, **execution stations** may be set up with real devices that are permanently connected by USB by keeping options such as auto unlock and never sleep enabled. Using USB connection also ensures that devices are always connected to the charger.

- With tools such as SeeTest that take up a lot of RAM while executing, it is recommended that the execution stations in the lab setup are kept with only up to 10 devices per station.

- As SeeTest is in turn triggered via the external execution environment, it is a best practice to launch it from the command line in order to avoid the floating license feature box. This can be done using the following command:

  `prompt:studio.exe -skip-features-choose`

- While connecting by USB, it is important to note that OEM-specific applications such as Samsung Kies or HTC Sync will try to kill the existing ADB process and use their built-in ADB. So it is better to avoid using those tools and only install the USB driver on execution stations.

- If the connected devices are not detected by the tools, then restarting the Windows service **IpOverUsbSvc** (Windows Phone IP over USB Transport) usually resolves the problem.

- For a iOS simulator with SeeTest, the simulator needs to be installed on a Mac machine using XCode, and when configuring the iOS device, the IP address of the Mac machine needs to be provided.

- With a BB emulator, there is no specific configuration required, but it needs to be set with 100 percent zoom and run without a skin, with the dim backlight option disabled so that it provides maximum compatibility with the real device.

- The iOS `.ipa` file that is used for instrumentation should not be downloaded from iTunes directly. Such files have DRM restrictions which will not allow the user to instrument the application. The application should be a build version without DRM restrictions, which can be obtained from the development team.

- For testing applications deployed in multiple languages, **SeeTestAutomation** supports specifying any given language properties file to be used during script execution. Thus, the same test script can be utilized to test in multiple languages.

# Summary

In this chapter, you learnt about automation techniques with real devices. By taking examples from both SeeTest and Appium, we have built a base that can be elaborated to other tools as well. Although different tools have individual differences, the techniques described are common to all. Now we will build on these concepts in the next chapter, in which we will learn about the automation techniques with cloud-connected devices.

# 6

# Automating on Cloud

In this chapter, we will learn the techniques required to automate a mobile application with cloud devices. This approach is the most advanced and user-friendly way of automation, although, at the same time, it requires a lot of investment on infrastructure and tool licenses. Since in this approach, too, we are working with real devices, it is equally reliable as the real device-based automation testing. Most cloud tools also provide a web-based interface or a thin client, which can be used to connect to real devices remotely from any given location. In this chapter, we will learn the various tools that can be used for automation in cloud setup and the process with which these tools are required to be set up, and also learn the most effective ways to automate with this technique. While understanding the implementation, we will also learn the real-world practical limitations of this technique and some ways in which to overcome these shortcomings.

We will cover the following topics:

- Some prominent Mobile Cloud automation tools and their features
- Process to set up private cloud labs
- Code snippets with working examples
- Advantages and limitations of this technique
- Troubleshooting and best practices

## Getting started with cloud automation

A mobile cloud tool is a platform that allows testers from varying geographic locations to access and test real mobile devices spread over different locations via a network such as the Internet or intranet. Basically, a cloud tool provides real-time access to mobile devices connected to real mobile networks through a web interface or a thin client within client-server architecture.

In the following figure, you can see that with a cloud-based tool, different teams from various geographic locations can access the same set of devices, which can themselves be distributed in various geographic locations as per project needs. Since the devices are connected over the cloud, it is very simple to capture mobile application performance data also and, hence, these tools are also commonly used for performance testing along with mobile functional and automation tools.

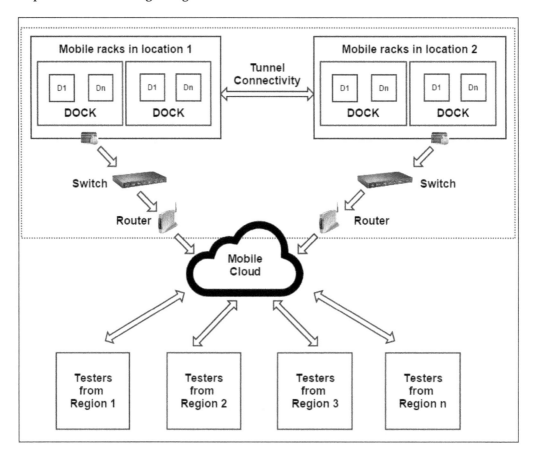

Devices are connected with device racks that contain device cradles. A commonly used device cradle is shown in the following figure. Each rack can support multiple cradle devices and some can support up to 20 devices simultaneously. With this kind of hardware, all controls such as audio, screen interactions, and even hardware buttons may be controlled remotely via the Web or thin client interface.

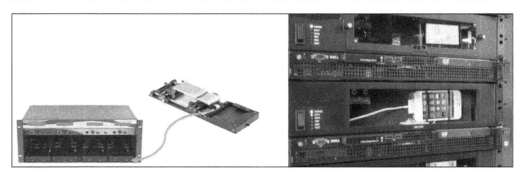

There are various types of cloud services, let's take a look at some of them:

- **Public cloud setup**: It is a publicly hosted service from which organizations can buy a certain amount of usage and can engage in a **pay-as-you-go** model. So, these are licensed software and device usage plans have to be purchased from the service provider. The device usage plans that vary from region to region, but are typically in the range of 30–50 dollars per device per hour. The cost of the hourly usage may also vary from the kind of services that are provided, such as functional testing only, or a complete package with functional automation and performance testing capabilities.

- **Dedicated private cloud setup**: An in-premise cloud setup may be done using hardware that is procured from a service provider and devices that are owned by a company. In this case, various teams within the company may share the same infrastructure in a time-bound manner and thus it proves to be more cost effective in the long run. Organizations that are security conscious, especially the ones that deal with sensitive data that is regulated such as insurance, healthcare, and banking, tend to go for this kind of infrastructure setup.

- **Shared private cloud setup**: A third party may use the infrastructure setup of a cloud service provider to set up a cloud service that can then be used by different customers. Large IT services companies usually employ this setup because it relieves their customers from the hassles of procuring devices and setting up secure facilities. This setup provides security as well as ease of setup, but at a slightly higher cost than a dedicated private setup.

So, even though cloud-based mobile testing solutions might sound like a buzzword, they are a very effective technique to enable mobile automation with real devices, especially with teams that are geographically distributed or have a global delivery model that is an offshore-onsite model.

Because of the flexibility and security they provide, cloud tools are most prominently used for large engagements and customers looking for enterprise-level solutions.

# Prominent mobile cloud automation tools

There are a large number of mobile cloud automation tools available in the market. Some of these integrate with other third-party functional automation tools such as **UFT** and **Selenium** to enable and support automation. Some functional automation tools provide their own in-built automation mechanisms and some also provide a facility to host and set up a cloud lab. The following is a list of prominent mobile cloud tools that support automation:

- **Keynote Mobile testing (formerly DeviceAnywhere)**: This is the pioneer of Mobile Cloud testing solutions and provides coverage to a large range of devices, especially the legacy ones such as old Nokia and Blackberry smart phones. It also provides an automation solution, which has a comparatively higher license cost. For connection to devices, it provides both a thin client and a web interface. It provides solutions to integrate with HP UFT and ALM along with support for IBM RFT and Selenium. It also provides out-of-the-box support for the Jenkins continuous integration tool.

- **Perfecto Mobile**: This is a market leader in terms of Enterprise mobile testing solutions as it provides end-to-end services such as functional testing, performance data capture on devices, as well as an integrated automation solution. It also provides an automation plugin for tools such as UFT and Selenium. It provides both enterprise-level as well as private lab setup facilities. Its typical licensing is available both on the per hourly usage cost, as well as for a dedicated lab, either as a publicly-hosted secure lab or an in-premise lab setup.

- **TestPlant EggCloud**: Testplant's EggCloud is a service that is provided typically as an add-on to the EggPlant test automation tool. It is mainly used by customers who are already using EggPlant as an automation tool, which is an OCR-based automation framework.

- **SeeTest Cloud**: This is a cloud setup facility that is provided with SeeTest and it can be used to host a lab with real devices that can be remotely accessed. Since SeeTest is the primary tool, its automation facility is also tightly coupled with SeeTest as an automation solution, although for execution, any of the execution frameworks that SeeTest supports may be used.

- **Ranorex**: Ranorex provides a simple way to set up a remote access to devices through the Ranorex tool interface and can be used in conjunction with Ranorex automation tool for automation end-to-end test scenarios on both iOS and Android devices.

- **SOASTA**: This is a facility that is sold as a separate license and can be used for automation testing on real devices in conjunction with SOASTA tool. It supports object-based scripting, and even the scripts can be stored on cloud, which may be accessed through a web interface.

- **Mobile Labs**: This is one of the fastest-emerging mobile cloud providers, which is typically gaining acceptance with customers who prefer an in-premise cloud setup, since it provides a very easy-to-set-up **Mobile Cart** hardware, as well as its server infrastructure needs are also very limited. Due to this, the costs involved are relatively low. It provides both functional and automation testing capabilities and is primarily geared toward HP UFT as the automation tool, although it also provides an integration capability with Selenium WebDriver, but this is limited to only supporting mobile web applications.

As is common with all emerging technologies, mobile test automation is an area where lots of functional automation tools have recently started providing cloud support services, so one should always do a thorough market research before choosing any one cloud setup provider.

# A private cloud lab setup

Before we move on to learning about creation of automation scripts in this technique, it is very important to understand the procedure involved in setting up a mobile cloud lab. If your organization has taken a public cloud license service where the basic paying model is pay-per-use or pay-as-you-go, then the setup part is not included in the test automation teams' responsibilities. But even then, it is important to be aware of the mechanism. So, in this section, we will learn about the process to set up a private mobile cloud lab, using **Mobile Labs Device Connect** as an example. This is purely illustrative and may vary slightly from different tools' perspectives; however, the overall procedure remains the same.

 The hardware requirements vary depending on different tools and the number of devices to be supported with the cloud setup. This information is usually provided with the installation guide documents.

First of all, we need to install the required software on the server machine, which will act as the host and on which all devices will be connected to set up the cloud lab. For this, Mobile Lab provides a Mobile Cart, as shown in the following image, which has a self-contained physical unit to set up the device lab. It contains a device rack with a capacity to connect up to 48 devices over three parallel **Datamation** 16-port USB Charger and Sync hubs.

The server machine is an Apple Mac Mini running Mac OSX 10.9.5 or higher. Other configuration parameters are 16 GB RAM and 1 TB hard disk space with Intel i7 2.3 GHz processor.

After connecting the required devices in the Mobile Cart, we need to configure them and the application to be tested, which is explained in the following step-by-step guide:

1. **Configuring Mobile devices**: As we learnt in the previous chapters, we need to disable the screen auto lock and USB debugging features on the devices. Along with these, we need to make the following configuration changes on devices:

   ° **iOS Devices**: Accessibility options on iOS need to be enabled for allowing data connection and device input via computer keyboard entry. For iOS 8 devices, the UI automation option also needs to be enabled.

   ° For iOS 7 and above, a **Trust This Computer?** dialog will be displayed on the phone after connecting the device to the server. Click on the **Trust** button on the device to enable the device control.

- ° **Android Devices**: Turn on the **Developer** options from the following path: **Settings | Applications | Development**.

- ° Check **Allow Mock Locations**.

- ° Enable the Checkbox in the **RSA Key** dialog to **always trust the computer**.

2. **Provisioning deviceControl applications on connected devices**: For Android devices, there is no specific provisioning required; however, for iOS devices, there is a need to provision the **deviceControl** applications.

   - ° **iOS Devices**: After the server setup and device configuration is complete, navigate to the `http://<<server IP address>>/` and log in with the default admin user. Navigate to the **System** tab and download the iOS provisioning Tool.

   - ° Choose an iOS developer profile and run the tool for the **Trust Browser**, **deviceControl**, and **deviceControl** iOS5 applications

   - ° Click on the **Continue** button and select a location to save the provisioned applications in the **codesign wants to use your key** dialog.

   - ° Upload the provisioned **deviceControl** applications in the **iOS Management** tab under the **System** tab, as shown in the following screenshot:

3. **Android Devices**: Although there is no additional configuration required for Android devices, in order to enable the **deviceViewer** application and the **Trust UFT** add-in, the application under test needs to have the INTERNET permission assigned in the manifest.xml file and the isolated process flag should be set to FALSE, as shown in the following code:

```
<uses-permission android:name="android.permission.INTERNET">
</uses-permission>
<service android:isolatedProcess="false" >
</service>
```

All connected devices will now be available under the **Devices** tab of **Mobile Lab**, as shown in the following screenshot. Devices that are ready to use are shown with a green icon:

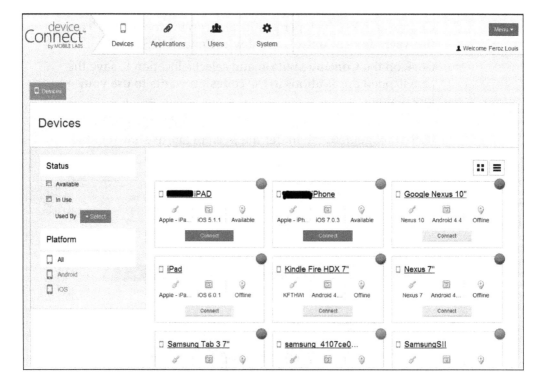

4. **User Management**: Now, create new users and provide the required users with admin privileges from the **Users** tab. Once set up, other users with admin privileges may be used rather than the default user to perform administrator functions, as shown in the following screenshot:

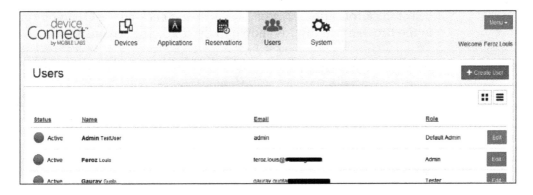

5. **AUT Management**: Under the **Applications** tab, all instrumented applications that are provided by default with Mobile Labs are listed for the first time. To instrument and upload any new application, use the **+ Add Files...** button, as shown in the following screenshot:

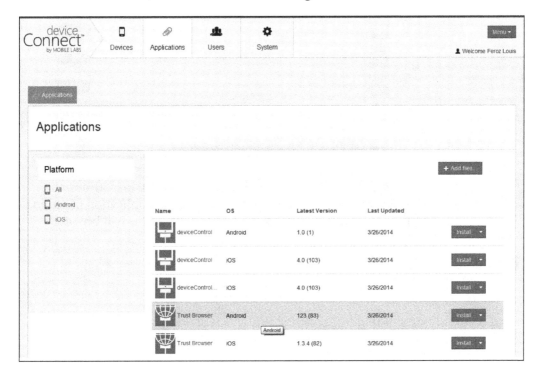

    °    Select the application `.apk` or `.ipa` file, and click on the **Open** button in the **Browse** window, as shown in the following screenshot:

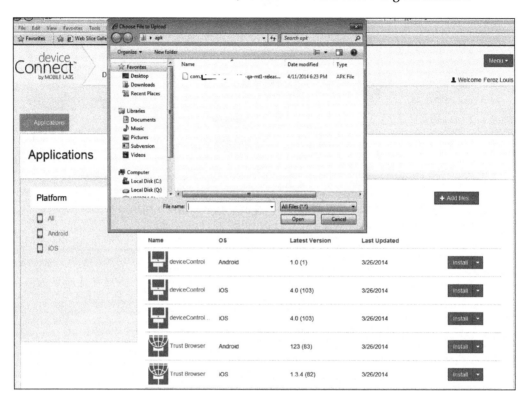

    °    During the upload process, Mobile Labs would also instrument it automatically to enable automation. When the application file upload completes, a success message will be shown, and it will become available in the application list.

6. **AUT installation on devices**: Mobile Labs allows for installation of applications on specific devices in one go. It allows for installing .apk files on Android and .ipa files on iOS. To install, select the .apk file from the application list and click on the **Install** button associated with the file. It will display all the available Android devices with the options **Install** and **Preserve Data**. By default, the **No** option is selected. If the user sets the **Install** option as **Yes** and **Preserve Data** option as **Yes**, then it will install the newer version of the application to the selected devices and preserve the data of previous versions of the same application, as shown in the following screenshot:

   ○   Select the options, as shown in the following screenshot, and click on the **Continue** button:

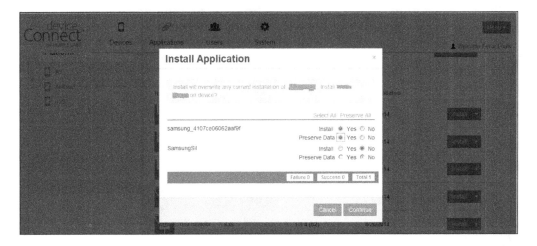

○ After processing, it will display the installation status with a green tick mark against the selected devices, if successful, as shown in the following screenshot. Click on the **Finish** button to complete the procedure. In this way, it is possible to install the same application on multiple devices in one go.

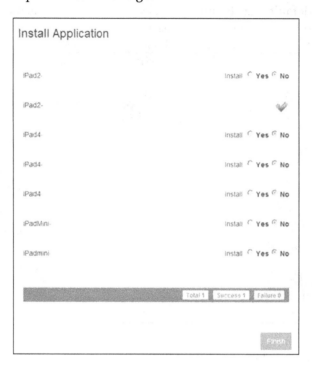

7. **Device Reservations**: Since, on a cloud network, multiple users are expected to share the same set of devices, a mechanism to manage time slots and devices is usually provided on all cloud tools. In Mobile Labs, this is provided as the **Reservations** tab, as shown in the following screenshot:

    °   Using the **+ Create** button under this tab, various users can book devices on a first come, first served basis by creating bookings. If a user closes a device, there is a prompt to release the device for other users, in which case it becomes available for others to use.

# Integrating automation tools such as UFT and Selenium

Now that we have understood the usage of cloud-based mobile testing tools—setting up the lab, connecting new devices, installing and instrumenting new mobile applications for functional and automation testing, and so on—we will now learn automation testing.

Some cloud tools such as **SOASTA** and **Perfecto Mobile** provide an inbuilt mechanism to automate test cases on cloud devices. These tools normally do not need any additional setup for automation, and the scripts are usually present on the cloud, which can be accessed remotely via the Web interface. However, most would require that an additional license be installed along with the cloud tool to enable automation. It is not necessary that all users be granted privilege to allow access to automation, and during user setup, this can be defined so that only a limited number of users can access and modify automation scripts.

Along with the inbuilt automation engine that is present with some tools, most of these tools also allow integration with external execution environments or tools such as UFT and Selenium. In this section, as an example, we will learn about the process to enable automation with UFT in Mobile Labs.

Although for manual testing Mobile Labs allows a pure web browser (Chrome and Firefox) based interface where the mobile screen loads in a browser pop-up window, for automation we first need to install the **deviceConnectCLI** and then also enable the **Mobile Labs Trust** add-in UFT/QTP. The download is available from the **Menu** option and needs administrator rights to be installed, as shown in the following screenshot. This enables a new add-in with the UFT/QTP interface and can be selected while launching the tool. When loaded, it recognizes the **deviceConnectCLI** as a mobile device and all internal controls of the AUT can be extracted and saved in the QTP/UFT object repository.

Most of the advanced functions such as `GetROProperty` and `SetTOProperty` work properly. However, some of the features such as descriptive programming may or may not work properly, depending on the type of object.

Similarly, Experitest SeeTest provides an add-on (`http://d242m5chux1g9j.` `cloudfront.net/SeeTest_windows_8_0_47.exe` or `http://experitest.com/` `download/seetestautomation/`) that can be directly enabled in UFT to create automation scripts with real devices connected through the USB or with the SeeTest cloud service. For Selenium, the driver can be downloaded from `http://experitest.` `com/plug-insqtptestcomplete/plug-in-for-selenium/`.

The following screenshot shows the HP UFT tool's interface:

The most important advantage of the plugin or add-in is that, after installation, the UFT object spy can be directly used to identify object properties instead of relying on the tool interface. Let's take a look at the following screenshot:

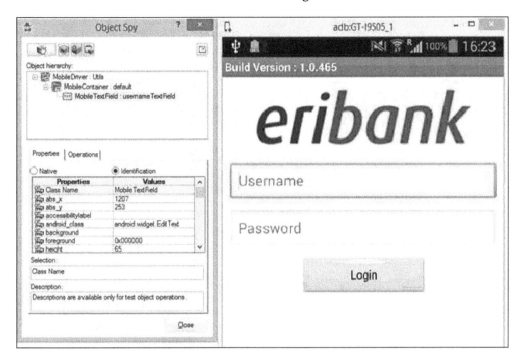

For example, after instrumenting and enabling the mobile automation add-in, the object properties can be viewed with UFT Object Spy. This makes the overall script development cycle much shorter, as there is no need to additionally export the recorded code to VBScript and all supported operations can also be viewed directly from the UFT object spy interface.

HP has partnered with Perfecto Mobile to develop an add-in called the **MobileCloud for HP**, which can be downloaded both from HP support and the Perfecto website (http://www.perfectomobile.com/ni/download-uft-addon). With the HP UFT Mobile Cloud add-on for Perfecto, the device screen loads directly into the UFT screen into a separate section. This makes the step recording very flexible and easy to work with. Using this add-on allows access to Perfecto-specific commands also, such as SMS me and Call me.

Let's take a look at the following screenshot:

It is important to note that although the basic mechanism is the same, the provider-specific commands and object class names might differ from one tool to another. So, it is necessary to refer to each provider's **Object Model Reference** before the start of scripting.

# Automation scripting with cloud devices

In a stepwise manner till now, we have understood how to set up devices and enable automation on the devices. Now we will take a look into automating mobile applications, first with the inbuilt automation engine of tools such as Perfecto Mobile and then with the automation tool-specific plugin such as UFT and Selenium.

# Automating with inbuilt cloud tool automation support

For this example, we will take Perfecto Mobile as the tool that provides a separate **Automation** tab. This tab would only be available for users that have logged in with user credentials that have been granted access to automation screen during user setup.

From this tab, various inbuilt keywords and object controls may be accessed. After selecting the **Automation** tab, the following screen is displayed:

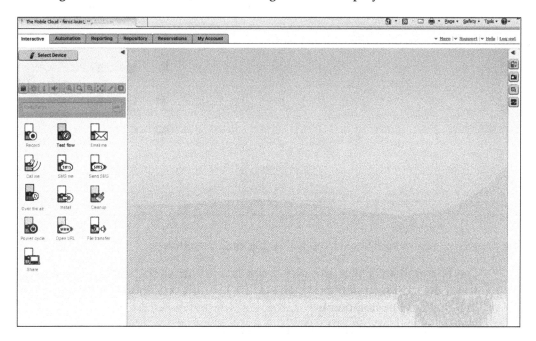

This screen doesn't provide any real-time code access and it only has a keyword display, which is also sometimes referred to as a **script-less automation** mechanism as it is not important to learn any tool-specific coding techniques or language to create automation scripts with this technique. Various application controls may be mapped to either image-based objects or to test objects, which are based on the actual application object properties as per the application code.

The following screenshot shows the Mobile Cloud tool's interface:

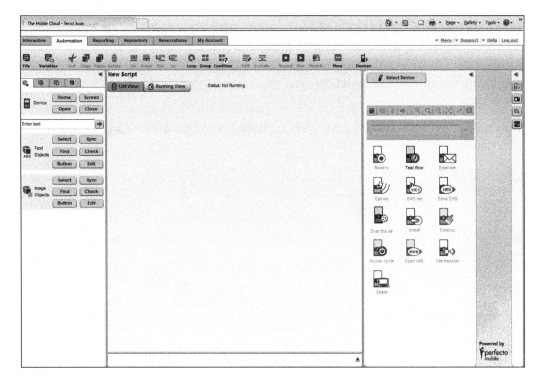

The preceding screenshot shows the following major sections:

- The **File** toolbar, which provides script-level controls such as creating a new script file, adding a data table, and adding conditional loops or grouping together a set of statements

- The second section, which provides statement support for **Device** controls such as **Home** key press and object-level controls, which allow adding **Test Objects** and **Image Objects** for object identification

- The third section, which displays the actual script statements and the various statements that make up the script and interactions with application objects and controls

- The fourth section, where device interaction controls are provided, such as **Select Device**, **Open URL**, or sending command to press the **Power Cycle** button

- The right-most section, where the device screen loads, which is used to create the recorded flow of scripts

To start creating an automation script, navigate to **File** menu and select the **Open** option shown with a folder icon, as shown in the following screenshot:

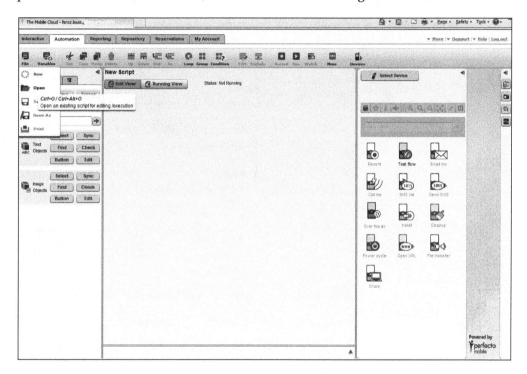

Various inbuilt functions may be accessed from this section. The commonly used functions are available with the first tab, and the complete list of functions may be accessed from the second tab. Specific functions are grouped together for easier access, for example, all device-related functions are present under the Device folder, as shown in the following screenshot:

The script section also allows opening of any previously created scripts or reusable functional components created. To open any previously created components, navigate to the My Scripts folder and then to the Supporting Functions folder, which will list all previously created components.

We can create the scripts directly by the recording of the flow on a device, which will be created in the high level programming language provided by Perfecto, having keywords for various user interactions such as **Select text**, **Browser.GoTo**, and **Edit.Set**.

The scripts can also be created by adding calls to components that were previously created, as shown in the following screenshot:

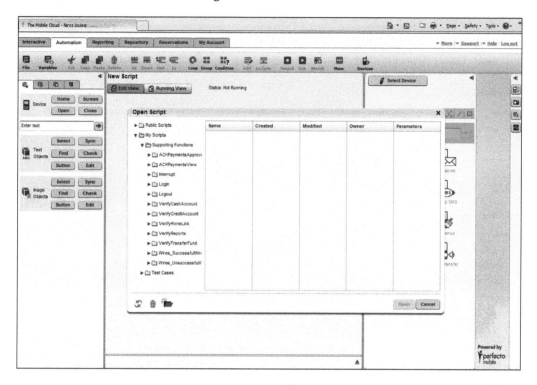

Perfecto also provides a virtual function mechanism, using which, these components can work independently of device screen or Operating System level changes. For example, a virtual function for **Login** can be created with individual sub-components for various OS and devices that differ in terms of the User Interface. So, there should be a component for classic devices like Blackberry and enhanced-look OS like iOS. Within an OS, there needs to be further sub division of device types, such as in iOS, there should be different components required for iPad and iPhone due to the difference in pixels on their respective screens, especially if image-based objects are used during the creation of script components.

The following screenshot shows the **Edit User Function** and the different options it contains:

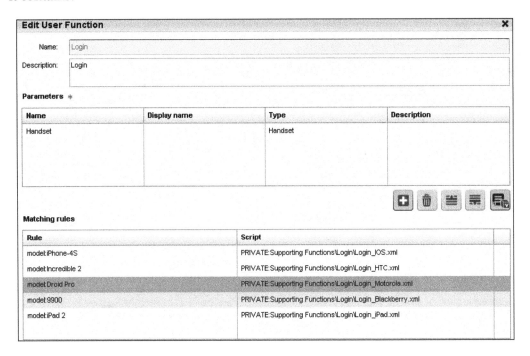

For the virtual function **Login**, separate subfunctions are called through a virtual function, which detects the DUT and picks the required function to execute. Using this functionality, scripts can be designed with only business logic implementation, and the internal mechanism to control execution for devices may be left encapsulated with the business function. This helps reduce the maintenance effort in case of changes to the application flow or specific modifications to a business function.

The following screenshot shows the Mobile Cloud tool's interface:

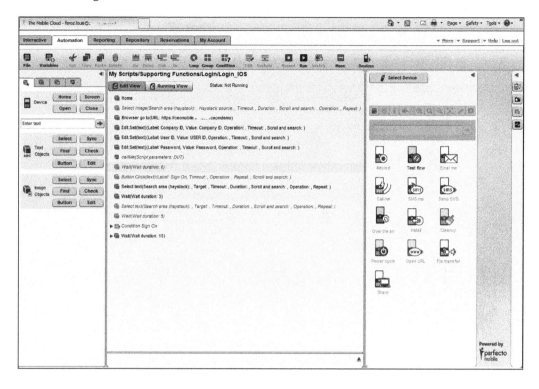

When we click on the **Devices** button, a window will open showing the various available devices on which the script recording (or execution during run session) may be carried out, as shown in the following screenshot. The device list shown depends on the currently available devices or the devices for which a reservation has been made with the User ID that is currently logged in.

Click on any available device and open it. At this step, the Perfecto server also internally checks if the account has some usage balance that can be utilized if a pay-as-you-go or public cloud service is being utilized. If sufficient balance is not present with the currently logged in user, then an error is shown. To create steps, objects need to be added and then used to perform actions.

As shown in the following screenshot, a basic flow of navigating to a URL within a mobile browser and then performing the login process is created. Each keyword such as `Edit.Set` would require an object to be added. The objects are embedded along with the step, and any further modifications need to be done at the step level, so although there is not much scripting expertise required with the inbuilt Perfecto engine, it is very important to create the objects very carefully.

As an additional step to verify object identification, it is advisable to test the objects on devices other than the original device on which the object was created.

To add steps such as validation of any condition or presence of any specific object on screen, the Condition keyword is used. Perfecto also provides a Wait method, which may be used to do synchronization of script steps. User gestures such as swipe and scroll may also be performed with specific keywords for this operation.

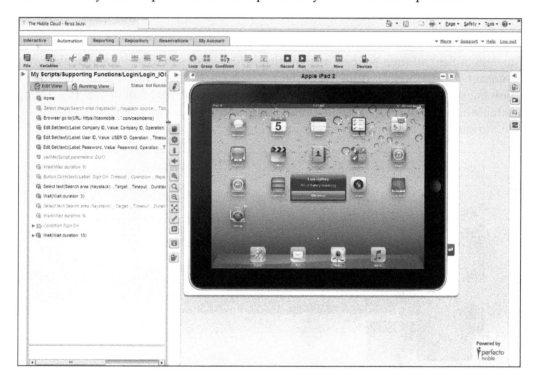

# Test data parameterization

The test data for scripts is kept separately in data tables that are present in the **Repository** tab. In this example, we will parameterize the device ID so that the same script may be executed over multiple devices in sequential execution mode. With this facility, Perfecto allows integrating the keywords with a data-driven methodology, thereby providing support for a hybrid framework.

To create a new data table, navigate to **My Data Tables** or **Public Data Tables**, and then click on the **Add** button to add new columns. To add values under each column, just make entries with the required data.

The data table thus created may be added as an input to the **Loop** command and values from its columns may be used during execution, as shown in the following screenshot:

The data tables can then be used with multiple test cases since they are available as reusable components within the Perfecto interface. In the example shown in the following screenshot, a data table has been made with entries of all device IDs. Using this table as input over the launch device component, the scripts can be executed over various devices in multiple iterations.

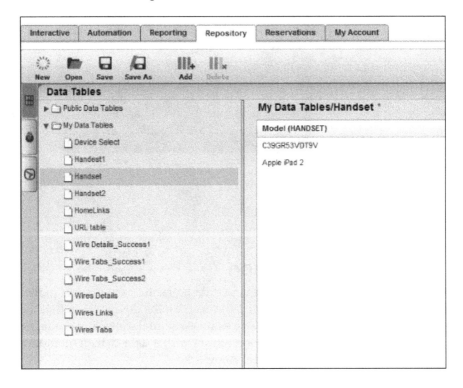

# Interrupt scenario automation

Perfecto Mobile also provides support for some very handy features that allow creating interrupt scenarios, such as getting a call during an ongoing transaction or getting an SMS while the active screen has the application under test loaded, as shown in the following screenshot.

Using the **Call Me** option, an incoming call can be made to coincide with a transaction step such as the login button press, which posts the data to the application server. Similarly, the **SMS Me** option simulates the receipt of an SMS during the transaction. Since the **Call Me** and **SMS Me** options are governed by the real network conditions, it is usually advisable to add sufficient delay, and to add the **Call Me** or **SMS Me** option a few steps before the actual transaction step.

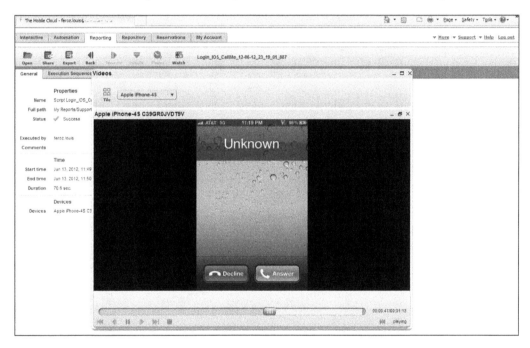

# Execution and reporting

To replay the execution of a recorded session run previously, navigate to the **Reporting** section and click on **Open**.

In the **Select Report** pop-up, expand the My Reports folder and then go to the Test Cases folder. On the right-hand side, all previous runs will be listed and the video of execution can be played to understand the exact cause of failure, as shown in the following screenshot:

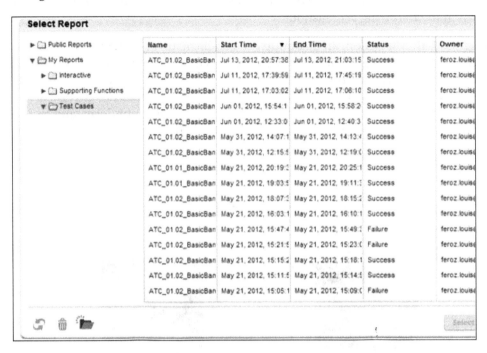

Double-click on any result and then use the **Watch Video** button to start the automatically captured execution video, as shown in the following screenshot:

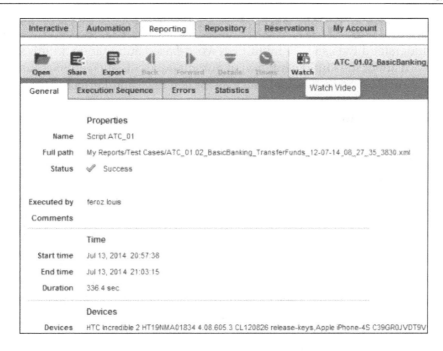

A unique feature with Perfect allows simultaneous viewing of multiple devices in a split window, which will open with a split pane of various devices with the stepwise continuous execution video of both devices, as shown in following screenshot:

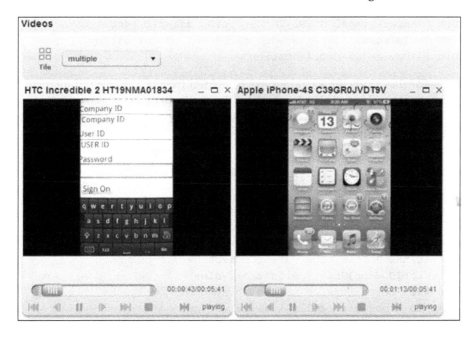

# Automating with third-party automation tools

Having understood the usage of inbuilt keyword-based automation, in this section, we will learn about the implementation of automation tools such as Selenium and UFT with mobile cloud testing tools such as Mobile Labs and Perfecto Mobile. Use of these tools instead of the inbuilt frameworks provides flexibility to implement an automation solution to work with a known tool and language. It is also easier to find experts in these tools, so the overall cost of implementation is reduced.

## Scripting with UFT

After enabling automation tools that we learned in the previous sections, in the following example code snippet, we will create a set of functions and then use these functions to create a script which will cover the end-to-end flow of the test scenario. The scripting is pretty much the same as with any other available add-in with UFT, so people who are well-versed in UFT scripting would just need to learn the basic object model and the process with which to trigger mobile devices through **VBScript**. For these, various tools provide predefined functions and the help document comes in handy to identify relevant code snippets. However, it is equally important to implement the functions in a proper way so that the execution and maintenance effort is reduced. The following implementation is primarily a modular framework with mobile device data being read from an external input sheet:

```
Function fn_SelectDevice()
   Dim DeviceName, OSVersion, DeviceType, ApplicationType
   Dim connectParams, dcURL, dcDeviceID, dcAppID, dcUsername
   Dim dcPassword, dcScales, dcAutoconnect, dcInstallApp,
   Dim i, DeviceID, dcOrientation
```

The values for these specific variables are imported from an external data input sheet. Use of external input sheets provides the flexibility to execute scripts over various devices without needing to modify actual script code. Let's take a look at the following code snippet:

```
   DeviceName = objSheet.cells(i,2)
   OSVersion = objSheet.cells(i,4)
   DeviceType = objSheet.cells(i,3)
   ApplicationType = objSheet.cells(i,5)
'Get Connection Paramerts and build Connection string
   dcURL = objSheet.cells(i,6).Value
'This is the Mobile Labs deviceConnect URL/IP and port
   dcDeviceID = objSheet.cells(i,7).Value
'this will be the device name displayed on MobileLabs interface
   dcAppID = objSheet.cells(i,8).Value
```

```
'This will be App Name displayed on Mobile Labs
  dcUsername =objSheet.cells(i,9).Value
'this is the required Mobile Labs deviceConnect username
  dcPassword =objSheet.cells(i,10).Value
'this is the required Mobile Labs devicConnect password
  dcScales = objSheet.cells(i,11).Value
'this is the scale you would like to set
  dcAutoconnect = objSheet.cells(i,12).Value
'this is required to be set as "True" if you want to autoconnect
  dcInstallApp= objSheet.cells(i,13).Value
'this is required to set as "True" if a fresh app install is
required
  dcOrientation = objSheet.cells(i,14).Value
'this is required to launch the device in portrait of landscape
mode
  connectParams = dcURL _
  & " " & dcUsername _
  & " " & dcPassword _
  & " -device " & dcDeviceID _
  & " -scale " & dcScales _
  & " -orientation " & dcOrientation _
  & " -r " & chr(34) & dcAppID & chr(34)
```

Setting the device-specific information into environment variables, as shown in the following, allows for using them across the entire suite and not just within the function:

```
Environment.Value("DeviceName")= DeviceName
Environment.Value("DeviceOSVersion") = OSVersion
Environment.Value("DeviceType") = DeviceType
Environment.Value("ApplicationType") = ApplicationType
```

Closing an existing MobileLabs device Viewer is important as UFT or QTP can only work with one instance at a time due to their limitations. Let's take a look at the following code snippet:

```
SystemUtil.CloseProcessByName "MobileLabs.deviceViewer.exe"
Wait(7)
Reporter.ReportEvent micDone, "ConnectToDevice", "Connect
Parameters: " & connectParams
SystemUtil.Run "MobileLabs.DeviceConnect.Cli.exe",
connectParams, "C:\Program Files (x86)\Mobile Labs\Cli"
Wait(30)
End Function
```

In the launched device, using the `fn_openApp()` function, we will open a mobile browser and navigate to the e-mail application URL, as shown in the following:

```
Function fn_openApp()
MobiDevice("MobAndroid").MobiBrowser("browser").MobileEdit
("Address Bar").Set "http://www.testingsite.com"
MobiDevice("MobAndroid").MobiBrowser("browser").MobiButton("Go")
.Click
End Function
```

To log in, we will use the `fn_EnterDetails()` function, as shown in the following:

```
'Enter the details
Function fn_EnterDetails()
MobiDevice("MobAndroid").MobiBrowser("browser").MobileEdit("First
Name").Set "TestUserID1"
MobiDevice("MobAndroid").MobiBrowser("browser").MobileEdit("Last
Name").Set "TestUserPwd123"
End Function
```

The `fn_SubmitToCreate()` function will be used to submit the entered values to create the user:

```
Function fn_SubmitToCreate()
MobiDevice("MobAndroid").MobiBrowser("browser").MobiButton("Next
Step").click
End Function
```

Now, the scripts would be created by calling these various steps as functions from the main script body, as shown in the following:

```
'Script 1
Function TestScript01()
Call fn_SelectDevice()
Call fn_openApp()
Call fn_EnterDetails()
Call fn_SubmitToCreate()
End Function
```

# Scripting with Selenium

Many tools provide support with Selenium along with UFT because customers would want to continue using Selenium as their preferred tool for mobile automation if the functional automation team were already using it so as to bring uniformity of tool implementation and usage of automation frameworks.

In the following example, we will understand the usage of Selenium Web Driver with Mobile Labs connected devices through Android Driver that can be used to control a web browser within the launched device. For this, Mobile Labs provide a **Device Connect CLI** facility which can be invoked through the command-line interface.

```
public class BusinessComponents {
  void launchApp() {
    String url = Datatable.getColummValue("dcUrl");
    String username = Datatable.getColummValue("dcUserName");
    String password = Datatable.getColummValue("dcPassword");
    String deviceId = Datatable.getColummValue("dcDeviceId");
    String scale = Datatable.getColummValue("dcScale");
    String orientation =
    Datatable.getColummValue("dcOrientation");
    String appID = Datatable.getColummValue("dcAppId");
    String command = "MobileLabs.deviceViewer.exe
    "+url+""+username+""+password+""+deviceId+""+scale+"
    "+orientation+""+appID;
    Runtime r = Runtime.getRuntime();
    Process p = r.exec(command);
    DesiredCapabilities cap = DesiredCapabilities.AndroidDriver();
    cap.setCapability("version","4.4");
    cap.setCapability("platform","Anndroid");
    WebDriver driver = new RemoteWebDriver(new
    URL("http://localhost:8089/wd/hub"),cap);
    ad.get("http://www.bankingsite.com");
    Report.LogInfo("App launch","application launched
    successfully", Status.Done);
  }
```

The login() function is used to enter the user ID and password and perform login action, as shown in the following:

```
  void login() {
    String strUsername = Datatable.getColummValue("Userid")
    String strPassword = Datatable.getColummValue("Password")
    ad.findElement(By.name("UserId")).sendKeys(strUsername);
```

```
    ad.findElement(By.name("Password")).sendKeys(strPassword);
    ad.findElement(By.name("Submit")).click();
}
```

The `accountBalanceCheck()` function is used to validate if the balance is 10,000:

```
void accountBalanceCheck() {
    String accountBalance =sd.findElement
    (By.name("accountSummary")).getText();
    if(accountBalance.equals("10000")) {
        Report.LogInfo("Balance Check","Account Balance is 10000",
        Status.Pass);
    }
    else {
        Report.LogInfo("Balance Check","Account Balance is not
        10000",Status.Fail);
    }
}
```

The `logout()` function signs out of the browser session:

```
void logout() {
    ad.findElement(By.name("logout")).click();
    Report.LogInfo("Logout","Logged out
    successfully",Status.Done);
  }
}
```

Using these business components with a basic keyword, or a table-driven framework, or a hybrid automation framework, various scripts may be created. Since the launch application function has device ID and application ID parameterized, the same components may be called to create multiple automation scripts with variable data.

Perfecto Mobile provides the Selenium mobile driver and also an add-in for Eclipse, which can be configured directly in the IDE. This provides a full-fledged mobile scripting IDE with access to device screen and controls, as shown in the following screenshot:

With Java, it is also possible to run scripts in parallel using multi-threading concept, such as follows:

```java
import java.util.List;
import java.util.concurrent.ExecutorService;
import java.util.concurrent.Executors;
public class ParallelExecutionTest {
  public static void main(String[] args) {
    List<AppTestDeviceData> data =
    DemoAppTestDataProvider.getData();
    ExecutorService executor =
    Executors.newFixedThreadPool(data.size());
    for (DemoAppTestDeviceData deviceData : data) {
      MobileTest test = new MobileTest(deviceData);
      executor.execute(test);
    }
    executor.shutdown();
  }
}
```

Then, in your test, you call the `run()` method from the `execute()` method to run scripts in parallel over multiple devices, such as follows:

```
public MobileTest(DemoAppTestDeviceData deviceData) {
  _driver = new MobileDriver();
  _deviceData = deviceData;
  _device = _driver.getDevice(deviceData.getDeviceId());
}
public void execute() {
  run();
}
```

# Troubleshooting and best practices

Let's take a look at the following troubleshooting and best practices that we can follow:

- It is important to configure image-recognition-based objects with enough tolerance so that the same object may work with various screen resolutions and device operating systems. For this, many tools allow the provision of values as percentages rather than absolute pixels, and this improves the object recognition. Let's take a look at the following screenshot:

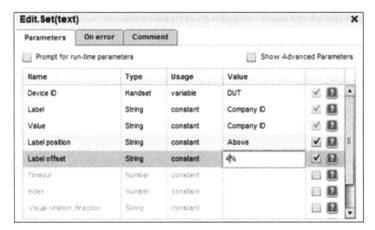

- Making a DNS entry for the mobile cloud server machine makes server search easier for end users. For example, `testMobileLabs.myComany.co.in` would make sure that navigating to `http://testMobileLabs.myCompany.co.in/Login` would redirect to the Mobile Labs server page regardless of the IP address.

- If the devices being used for testing are enrolled in a **Mobile Device Management** (**MDM**) program, there may be some policies that are incompatible with the use of cloud tools like Mobile Labs **deviceConnect**, which get installed on the device. So, it is best to ask for exemption for these devices because such devices are always accessed by users that are authenticated to the server, meaning that the security of the test devices will remain high.

- Any security policy that requires a PIN or password must be removed from the test devices connected on cloud tools as these PIN and password locks prevent devices from connecting to any computer until the device is unlocked.

- Tools such as UFT/QTP/TestComplete that use scripting languages such as VBScript can only access one target for a script. So, it is important to close any extra device windows prior to the start of execution. This should ideally be done programmatically in the launch device or setup functions.

- When working with objects that do not have an ID property value, use ordinal identifiers in order to uniquely identify each object.

- With UFT/QTP, an important best practice is to manually define a Page object in the **Object Repository Manager** (**Object | Define New Test Object**), and then cut and paste the children under that Page object. This makes the OR management much more simple and maintainable.

- Sometimes with mobile tools, UFT will allow accessing an object that is not in view but on the same page (by scrolling down) during web-based automated testing. However, sometimes such lines of code may malfunction and clicks might fail. It is recommended that as best practice, for such objects, the scroll method is used so as to have the object in view, ensuring that the item is accessible at the time of the click event.

# Summary

In this chapter, you learned about the use of cloud tools for mobile test automation. And with this, you have understood all the various device automation techniques. In the next chapter, we will take up using these techniques in real-time projects with a combined approach so as to have the most optimized solution.

# 7
# Optimizing Test Strategy and Estimation

Test strategy and effective estimation are the cornerstones of the success of any mobile testing endeavor. Keeping this in mind, in this chapter, you will learn about the processes and methodologies to define an effective mobile testing automation strategy that caters to the demands of any specific mobile assignment. We will discuss the factors that should be considered while optimizing the test strategy and how to effectively determine the optimization of the test strategy. You will also learn the estimation techniques involved in a mobile functional and automation projects, as both of these go hand in hand. Not only this, you will also learn how to predict the **return on investment (RoI)** of a project with various methodologies, so that an informed choice can be made right at the start of a project about the methodology and approach to be selected, and a suitable test automation strategy could be devised to attain the same.

As we proceed in this chapter, you will learn how to develop the most suitable strategy that focuses on maximizing the RoI and the ways to avoid pitfalls.

In this chapter, you will understand how to optimize a project's mobile test automation strategy when we cover the following topics:

- **Software Testing Lifecycle (STLC)** of a mobile test automation project
- Understanding the parameters to be considered to select the best suited approach for any project
- Estimating the mobile automation effort
- Maximizing the return on investment

# Lifecycle of a mobile test automation project

Project managers realize that an effective test strategy is the most important aspect of any project, but a mobile testing project cannot be treated in the same way as any other functional testing project. In this section, you will understand the nuances that differentiate a mobile testing project from a usual functional testing project in the various lifecycle stages of mobile test automation. The following are the stages that are followed in a mobile test automation project:

- **The planning stages**: These steps are usually taken to effectively plan a project. These are very important and are required to be performed before proceeding to the actual automation. The various steps covered are:

  1. Requirement gathering and analysis.
  2. Device matrix creation.
  3. Test automation strategy.
  4. Automation feasibility analysis.
  5. Effort estimation.
  6. RoI analysis.

- **The execution stages**: These steps are for the design, development, and execution of the automation test suite. The various steps covered are:

  1. Automation framework creation.
  2. Test scripting.
  3. Execution.
  4. Maintenance of automation scripts for new features in AUT.

We have already covered the automation framework creation and test scripting aspects in the previous chapters. In the next sections of this chapter, we will look at the project planning aspects pertaining to the relevant project lifecycle stages and steps with respect to the optimization of the test strategy and efficient project execution. This systematic approach to mobile automation projects helps to ensure maximum and earliest possible return on investment.

# Requirement gathering and analysis

Right from the first stage of requirement gathering and analysis, you have to differentiate the approach to plan and execute the project. In most cases, an organization ventures to extend its existing application to the mobile channel of access, so as to reach more end users and thus expand their user base. So, it is imperative that the testing team has a thorough understanding of the type of application the customer is targeting. If it happens to be a web application, then getting to know the supported set of browsers right at the beginning is important for proper planning.

Having a thorough understanding helps both development and testing teams to target the right operating system browser combinations for testing. However, while gathering this information, it is also equally important to understand that 100 percent testing coverage can never be achieved. So, information that can be used for prioritization should also be gathered.

The following is an ideal stepwise approach to identifying the requirements in a mobile testing project:

- **Identify the type of application that the customer wants**: If the application is a web application, then it is important to understand whether it will be built on top of the existing application infrastructure by using the same database schema and tables or it will be kept separate. This would help during the test design phase. For a hybrid application, it is important to understand the amount of reusable code that is going to be common between the various supported operating systems. This information helps to identify the operating system-specific test cases for the application and clearly defines how much of the application functionality can be commonly tested across different operating systems.

- **Understanding the application architecture**: In some cases, the organization launches both a mobile web version as well as a hybrid version of the application. In these cases, both the applications should ideally be tested as independent applications and separate testing cycles should be carried out. To design an optimum automated regression suite, it is good to understand how much of the functionality is common and how much of the code is shared between the mobile web and hybrid versions. With this information available, the regression testing suite can be devised to work on both the web as well as the hybrid versions, by making use of regular expressions in the object descriptions. For manual testing, this information also helps to optimize the execution spread across different devices.

- **Identifying the nonfunctional requirements**: Nonfunctional testing is more significant on mobile testing projects because it is not only important to know the kind of user volume an application is going to support, that is, server-side testing, but it is also very important to understand the resources that the application would require on the devices that are going to be supported. Identifying this performance engineering information also helps the development team to devise suitable strategies to support devices and keep the resource usage within acceptable limits. Factors such as battery consumption and network bandwidth usage should also be considered along with these requirements.

- **Identifying the form factors to be supported**: Normally, the mobile applications are created separately for tablets and smart phones, but with the increasing adaptation of responsive web design techniques, this boundary is being blurred. For testing teams, it is important to understand the amount of responsiveness that the application code can sustain. Thus, this information is important to define test scripts that test enough of the boundary conditions to guarantee defect-free application in production conditions. During this phase, it is also important to understand whether the application would support the transition between **portrait** and **landscape** modes, which later helps to define test cases to test such requirements.

- **Identifying application interruption behavior**: In some cases, the applications might be required to silently keep posting data on the servers, so that in the case of a catastrophic interrupt, such as the battery dying during an in-progress transaction, the information is not completely lost. The application dealing with financial data must have a clearly-defined interruption behavior, and testing teams should make sure to understand these requirements even before the start of testing, so that proper test cases to test such conditions are developed. Although most of these test cases are difficult to automate, but with cloud testing tools, many such interrupt scenarios can reliably be replicated.

- **Identifying information security needs**: Some applications store data locally on the devices that can be available to other programs. Although many mobile operating systems have intrinsic sandboxing capabilities, which limit inter-application access, given the potentially limitless permutations of operating system version and hardware combinations that a mobile application can be used with, it is important to understand the impact of these requirements on the functional design of the application as well. In most cases, such test cases are not required to be included as part of the regression suite, and hence, the testing team may not want to automate them. However, it is important to include a few basic checks, so that such a coverage is not totally missed in the final regression suite.

# Device matrix creation

For a mobile testing project, this is one of the most important stages and, in some cases, it is not considered as a separate stage but a part of the requirement analysis phase itself. A device matrix is generally required only when a high-level specification is provided at the start of the project about the type of application—native, hybrid, and web; the form factors—tablet, phones, and phablets; and the operating systems—Android, iOS, Windows, Tizen, Blackberry, and others are specified without providing the exact details regarding each of these. Regardless of the amount of details provided at the start of a project, it is imperative for a testing team to identify the exact devices and their operating systems' versions to provide optimum coverage during testing.

In this stage, the testing team can start the work of defining the spread of execution over various kinds of devices. Here, we recommend that you keep a few different categories of devices, such as:

- High priority / medium priority / low priority
- Must have / nice to have / no need to have
- Primary / secondary / tertiary
- Latest in use / most common in use / earliest in use

In the following table, we have a sample device selection matrix created based on the latest / most common, and oldest criteria. Similarly, for the other criteria also, we can create such a device coverage matrix.

It is very important to consider the device form factor in conjunction with the relevant criteria for the complete coverage of an operating system. The device matrix based on the latest / most common / earliest criteria are as follows:

| OS | Category | Phone | Tablet | Phablet |
|---|---|---|---|---|
| Android | Latest | S5 | Amazon Kindle Fire | Samsung Galaxy Note 4 |
| | Most common | Google Nexus | Galaxy Tab | Samsung Galaxy Note 2 |
| | Oldest | S2 | Motorola Xoom | Samsung Galaxy Note |
| iOS | Latest | iPhone 6 | iPad Air 2 | iPad Mini 3 |
| | Most common | iPhone 4S | iPad 3 | iPad Mini 2 |
| | Earliest | iPhone 3GS | iPad | iPad Mini |

| OS | Category | Phone | Tablet | Phablet |
|---|---|---|---|---|
| Windows | Latest | Nokia Lumia 920 | Acer Iconia | Lumia 1525 |
| | Most common | Nokia Lumia 530 | Windows Surface 2 | Lumia 1520 |
| | Earliest | HTC Windows Phone 8S | Windows Surface | |
| Blackberry | Latest | Q10 | Blackberry PlayBook | |
| | Most common | Z10 | | |
| | Earliest | BB Bold | | |

During the selection of devices for each of these categories, the following parameters need to be considered:

- The application's targeted geographic location and the expected device usage patterns for that geographic location.

- The operating system market share in the targeted geographic location or the general market trends for the operating system market share.

- Application types are native, hybrid, or web. Web applications also need to consider the various browsers and their market share of respective operating systems for complete coverage.

- Market trends and predictive analysis of the future user base. This is required to perform the optimum level of beta testing on the *soon-to-be-released* devices, so as to avoid untested bugs on a newly launched device.

- If this application is an extension of the preexisting application, then it can be based on the current usage analytics of the application, and the testing automation solution should closely try to mimic the device coverage from this device's analytics data from the production environment.

- An important factor to consider during device selection is the intended usage of the application and the degree of tolerance it can afford with respect to testing coverage. Financial-transactions-based applications have to be very thoroughly tested in order to avoid potential risk of untested bugs that might lead to loss of data or even fraud. In these cases, the nonfunctional requirements, especially the security testing, should be tested very thoroughly. Also, on the application level, all the devices should be covered as much as possible.

# Optimization of the device coverage matrix

After the identification of the devices to be covered in the project, the device selection should be optimized for efficient test planning and execution coverage, because in a normal project, it is practically impossible to test all the categories of devices. So, the testing team has to do some trade-offs with respect to device coverage in order to make sure that none of the devices remain completely untested, while, at the same time, we are not required to provide 100 percent coverage to each individual device. So, for example, if we have a total of 11 devices identified during the device selection stage for two different operating systems, such as Android and iOS, and two form factors, such as phone and tablet, in the previous stage, then the testing can be spread equally over each operating system and form factor, as shown in the following table:

| Device Name | Coverage |
|---|---|
| **Phone – Android** | |
| Samsung Galaxy S4 | 100 percent |
| LG Nexus 5 | 50 percent |
| HTC Desire HD | 50 percent |
| Phone - iOS | |
| iPhone 6 | 100 percent |
| iPhone 5C | 50 percent |
| iPhone 4S | 50percent |
| **Tablet – iOS (iPad)** | |
| iPad 2 | 50 percent |
| iPad Mini | 100 percent |
| iPad Air | 50 percent |
| **Tablet – Android** | |
| Samsun Galaxy Tab | 100 percent |
| Amazon Kindle Fire | 100 percent |

Table with device coverage of operating systems and form factors

This means that 100 percent of the test cases will be executed over the primary devices and only 50 percent of the test cases will be executed over the secondary devices. This execution spread has been designed to optimize the coverage from an execution perspective and to reduce the effort spent on redundant testing. However, this optimization should be done in a logical way. So, the reason to keep the coverage as 100 percent on both Galaxy Tab 2 and Kindle Fire is because, essentially, they are significantly different from each other and the risk of leaking a defect by having only 50 percent testing coverage on either of them is pretty high. By using this type of optimization, we can ensure that each operating system is tested for complete functionality and each form factor is tested for 100 percent of the functionality, by selecting the mutually exclusive test case sets of 50 percent each for the various categories of devices. These percentages can be increased or decreased as per the criticality of the application, the number of resources available, and the time available for test execution.

It is also advisable to use various connection mechanisms with the devices. So, some devices need to be set with the use of GSM, CDMA, LTE, and 4G, and some have to be set to use Wi-Fi. Within the GSM and CDMA devices, different network providers might be used to have maximum spread over network types and carriers.

If the AUT is a mobile web application, then the factor for native and third-party browsers should also be considered for device coverage. In such cases, the execution can be equally spread over each browser with only the GUI-specific test cases are to be repeated across each browser type.

 With automation, it might not be necessary to automate all the scripts for regression testing on all devices. This device optimization helps to reduce the effort for automation execution and analysis.

# Optimizing the test automation strategy

In the test automation strategy definition phase, the tool identification and acquiring process is carried out along with the identification of the automation approach to be selected. As we saw in the previous chapters, there are four major ways of automating any mobile application. So, at this stage of the project, the project manager and automation architect have to identify the most suitable approach to test the automation from the four ways, as follows:

- Automation using emulators and simulators
- Automation using user agents
- Automation using physically present real devices
- Automation using cloud devices

With each of these approaches, a variety of tools may be selected, which we discussed in the previous chapters, and their pros and cons should be thoroughly considered before finalizing any tool.

 Refer to *Chapter 2, Designing Mobile Automation Frameworks*, for the tool evaluation matrix. The parameters listed in *Chapter 2, Designing Mobile Automation Frameworks*, are exhaustive, and the relevant parameters should be evaluated while performing the tool evaluation exercise.

While selecting the automation approach, various trade-offs with respect to the application cost of quality may be considered. For example, for the given AUT, it may not be necessary to execute all the test scripts on real devices, and some test cases can be executed on simulators, emulators, or with user agents. This helps to bring down the cost of automation as user agents and emulators are free of cost, and thus, the overall cost of the usage of cloud devices or acquiring physical devices can be significantly reduced. With user agents, there is an added advantage that if a preexisting application is extended to the mobile platform that has a ready-to-use automated regression suite, then such an automation suite can be quickly modified to support user agents with relatively lower investment than creating a fresh automation suite to support the mobile web views of an application. This is generally possible because of the web page presentation technologies such as HTML and HTML 5; with basic modifications, the application can be made mobile enabled and, thus, for an existing automation suite, the object descriptions and properties remain unchanged or change very little.

With cloud-based services, there is no additional effort required to install and maintain device connectivity and security software. Automation with these tools is generally more cost-effective than using real devices, or even emulators or simulators, as most of the cloud tools provide access to the native application object properties and have extensive support. Also, there is no additional cost involved in acquiring real devices and keeping the database of the devices updated every quarter as new devices and operating system versions get released. However, during the set up of a private cloud lab, although the initial cost of the instruments and maintenance of devices is high, it is generally offset with multiplexed access of the same devices across time zones, across various teams, and to the people located at the low-cost development centers and offshore, which are generally more cost effective to operate.

A project manager has to take into consideration all these parameters and decide on the best-suited strategy for the project. A generally recommended approach is to execute the GUI-specific test cases exclusively on real or cloud devices and use emulators/simulators or user agents for the rest of the test cases.

For a typical mobile automation project, the distribution over emulators or user agents versus real or cloud devices can range from the ratio of 25:75 to 40:60 without any significant reduction in the defect detection efficiency of the automation suite.

It is generally recommended that the maximum amount of testing must be carried out on real devices. For the first round of testing, this is indeed the most suitable approach, but for test automation, which is generally more applicable for regression automation, it is not necessary to test the entire set of test cases on real devices. So, during the test automation strategy phase, project managers should make an informed choice about how much percentage of test cases they can safely execute over emulators, simulators, or user agents, which remains within the acceptable risk of defect leakage. This percentage can be easily determined by performing a high-level analysis of the mobile-device-specific requirements versus the total functional requirements. This percentage is generally a very good indicator of the amount of coverage that must be kept on a device and what can be moved to the simulated test bed.

In the test automation strategy phase, other assets required for the successful end-to-end delivery of the automation project, such as the hardware, software, coding guidelines, framework requirements, and human resources, are also identified. Usually, all these requirements are captured in a document called the **Test Strategy document** or, sometimes, it is included as part of the overall **Project Test Plan document**.

After the strategy phase, where the automation approach, distribution over combined approaches, software, hardware, and human resources for automation are defined, we move onto identification of automation-feasible test cases.

# The automation feasibility analysis

In this phase, we need to look at the type of test cases that are designed for the AUT and devices in scope. Since by this stage the tool and approach are generally identified, we need to focus only on the identification of test cases that need to be automated. During the tool identification application technology, support is verified and a tool that supports the majority of application controls is identified. However, there may still be some scenarios, such as interruption test cases or some specific controls, that are not feasible to automate, or test cases that can only executed one time so cannot be repeatedly executed, such as device activation over SMS. So, in this phase, such test cases are identified and removed from the automation-feasible suite.

The following parameters need to be considered while performing the automation feasibility analysis. This list can be adapted by the project under consideration; however, these are the most commonly used parameters for the automation feasibility analysis:

- The test case should not have any steps that are not supported for the automation or repeatable execution, such as one-time SMS-based confirmation, **one-time password** (**OTP**), or battery dead scenarios

- All the controls required to be used in the test case are supported with the automation tool

- The test data required with the test case is provided, which can be repeatedly reused, or precondition setup steps are provided, which can be automated, so that automation scripts would create their own test data

- Any validation steps that require access to third-party application interfaces or databases are verified to be automatable with the selected tool, and the required database queries are executable programmatically

To carry this analysis systematically, a checklist mechanism, as shown in the following table, can be employed. We have taken a sample project to highlight the usage of such a checklist:

| TC ID | TC description | All controls/ steps can be automated (Yes / No) | Required test data available or preconditions supplied to create required data | Access to all interfaces like external applications or databases provided (Yes / No) | Feasible/ not feasible |
|-------|----------------|----------------------------------------------|-------------------------------------------------------------------------------|-------------------------------------------------------------------------------------|------------------------|
| TC_01 | Account creation | Yes | Yes | Yes | Feasible |
| TC_02 | Eligibility check | Yes | No | Yes | Not feasible |
| TC_03 | Portrait/ Landscape toggle | Yes | Yes | Yes | Feasible |
| TC_04 | Claim creation | Yes | Yes | No | Not feasible |
| TC_05 | Insurance premium payment | Yes | Yes | Yes | Feasible |

| TC ID | TC description | All controls/ steps can be automated (Yes / No) | Required test data available or preconditions supplied to create required data | Access to all interfaces like external applications or databases provided (Yes / No) | Feasible/ not feasible |
|---|---|---|---|---|---|
| TC_06 | Insurance premium reminder | No (reminder dialog box is not identified with automation tool) | Yes | No | Not feasible |
| TC_07 | Policy registration | No (SMS OTP cannot be automated) | Yes | No | Not feasible |

Automation Feasibility Analysis based on checklist

After this stage, we will get clarity on the number of test cases that need to be automated. However, it is also recommended that besides the technical feasibility analysis, another stage of identification of test cases that are eligible to be included in the regression test suite should be carried out. In this phase, typically, test cases that are of higher priority and are more likely to result into defects are identified.

Even if there are enough funds available to the organization to automate all the automation-feasible test cases, the functional feasibility analysis phase helps to identify test cases that can be included as part of smoke, sanity, and regression suites. If permitted, further prioritization may be done within the regression test suite as primary and secondary, and the execution may be done on a periodic basis, so as to reduce the effort spent on the execution setup and analysis. The smoke and sanity packs can further be deployed with any continuous integration solutions, such as **Hudson** or **Jenkins**, for execution after every deployment, and the regression suite can be scheduled to run overnight every night at a predefined time.

# Effort estimation

After the feasibility analysis phase, a detailed estimation is required to be done so as to predict the effort required, and to identify whether the project is viable for automation and can provide a positive RoI. This effort estimation is the basic input to analyze the RoI for any project.

In this section, we will take a basic complexity-based estimation technique and learn how it can be utilized to estimate the effort made for design and execution activities on mobile devices. Along with the cost of this effort, the overall estimation will also include the cost of additional software licenses that may be required for mobile testing, such as a cloud tool, hosting server if a private cloud lab is to be set up, and of course the cost of devices. These one-time costs are in addition to the device usage costs that should be factored in if a pay-as-you-go model is being used, or if privately hosted devices have to use their own data plans and SIM cards.

To determine the overall size of a project, first of all, the complexity guidelines should be identified, and based, on the previous information available, such as the effort consumed in earlier projects, a base effort for each of the complexity types should be determined in as accurate way as possible. Since an estimate can only be close to the accurate and never provide an exact value, a general tolerance of 5-10 percent is considered normal. So, the estimated effort may vary by 10 percent above or below the actual effort consumed.

The following factors typically need to be considered while identifying the complexity of the expected test scripts that are to be developed:

- Number of steps in the test case
- Number of validations in the test case
- Number of transactions in the test case

For automation testing, generally, the manual test cases are readily available and these parameters can be accurately measured. However, if the test case documentation is not available readily, some assumptions need to be made by analyzing the requirements to determine the number of scripts for each complexity type as accurately as possible.

For each of the complexity types, these parameters should be assigned boundary values and the complexity should be defined very clearly in measurable terms with both upper and lower limits for all measurement parameters:

| Complexity type | No. of steps | Number of validations | Number of transactions |
|---|---|---|---|
| Simple | 1 to 10 | Less than 5 | Less than 3 |
| Medium | 11 to 35 | 5 to 10 | 3 to 6 |
| Complex | 36 to 50 | 11 to 25 | 7 to 9 |

Typical Complexity Guidelines in a mobile testing project

So, these values for any given test case can be used to define its complexity, and based on the available test case documentation, the test case count for each of the complexity types can be identified.

The total cost of the project should be identified by multiplying the base effort for each of the complexity types with the count of test cases identified for each complexity type. So, for a mobile native application that has 100 simple test cases, 250 medium-complexity test cases, and 125 complex test cases, the cost is calculated as shown in following table:

| Complexity type | Count of test cases [A] | Base design effort (person hours) [B] | Total Effort (person hours) =[A]*[B] |
|---|---|---|---|
| Simple | 100 | 4 | 400 |
| Medium | 250 | 8 | 2000 |
| Complex | 125 | 14 | 1750 |
| Total effort (person hours) | | | 4150 |

 The base design effort should be identified while considering the methodology to be used for automation. Typically, with open source tools and real devices, the base effort is higher as compared to COTS tools, such as HP UFT, and cloud tools, such as Mobile Labs and Perfecto Mobile.

Till this stage of estimation, total effort is pretty much the same as any typical test automation project. In many cases, this base effort is multiplied for each new mobile device. So, if there are six devices in scope, then the total effort is calculated as six times the base effort. However, this is not accurate if a proper framework is utilized for automation, which allows the reuse of objects across operating systems.

Hence, for proper estimation, the percentage reusability across different devices needs to be determined. For this, the information captured during the required gathering and analysis phase is very important. If the AUT happens to be a mobile web application, then the reusability is very high and device-specific customizations would be typically in the range of 5 percent to 10 percent only within the same form factor. Across the form factors, this range can be as high as 30 percent to 50 percent. For a purely native application, the code base is entirely different across different operating systems, but with newer technologies, there is still some amount of reusability. We will consider the case of a native application in an example now.

Let's continue now with the example taken earlier for the automation suite of a mobile native application that has the test case complexity distribution as **Simple: Medium: Complex::100: 250: 125** and requires 4150 person hours of effort for a base device. If this suit now needs to be extended to support a total of 11 devices, as considered earlier in the **Device Coverage Matrix** section, then the effort needs to be calculated while keeping in mind the modification effort required to adapt the scripts for additional devices. In this example, we are considering an automation tool that uses some amount of **Optical Character Recognition (OCR)**, due to which, such modifications might be required. The following table shows estimation for automation script design for all in scope devices:

| Device name | Modification effort (percentage) | Modification effort (person hours) | Remarks |
|---|---|---|---|
| **Phone – Android** | | | |
| Samsung Galaxy S4 | 100 percent (base device) | 4150 | Base effort |
| LG Nexus 5 | 15 percent | 4150*15% = 622.5 | This device has the same form factor as the base device, so 15 percent modification effort |
| HTC Desire HD | 15 percent | 4150*15% = 622.5 | |
| **Phone – iOS** | | | |
| iPhone 6 | 20 percent | 4150*20% = 830 | Since the first modification for the different operating system is the same for all of the form factor, hence the percentage is higher, up to 20 percent |
| iPhone 5C | 15 percent | 4150*15% = 622.5 | 15 percent modification effort for the same form factor after the first modification |
| iPhone 4S | 15 percent | 4150*15% = 622.5 | |
| **Tablet – iOS (iPad)** | | | |
| iPad 2 | 50 percent | 4150*50% = 2075 | 50 percent rework effort considered for the different form factor |
| iPad Mini | 15 percent | 4150*15% = 622.5 | 15 percent modification effort for the same operating system but slightly different form factor |
| iPad Air | 10 percent | 4150*10% = 415 | 10 percent rework required to make scripts work on the same operating system after the first modification |

| Tablet – Android | | | |
|---|---|---|---|
| Samsung Galaxy Tab | 15 percent | 4150*50% = 2075 | 15 percent modification effort for same operating system but different form factor |
| Amazon Kindle Fire | 10 percent | 4150*10% = 415 | 10 percent rework required to make scripts work on the same operating system with the same form factor after the first modification |
| Total effort | | 13072.5 | The total effort is calculated based on the consideration of the reusability across the operating system and form factors, and not a simple multiplication to the number of devices considered |

So, we saw that, even with an inefficient automation solution such as an OCR-based tool, the effort estimation is significantly different from the one calculated through multiplying by the number of devices. In the previous example, this value would have been 4175 * 11 = 45,925 person hours, which is very high as compared to 13,072.5 person hours.

For a web application with automation tools that use object properties for identification, we need not even consider specific devices individually during estimation. The effort can be calculated by considering the reusability across different operating systems. For some operating systems, the code base remains the same, and there is ideally no modification required unless there are any native operating-system-based objects to be used in the identification. For each operating system type, a modification effort of around 15 percent may be considered. So, for a project that has a base effort of 1000 person hours for a mobile web application that needs to be executed across mobile devices on four different operating systems, the total effort and the estimation for automation script design for all in-scope Mobile Operating System would be calculated as shown in the following table:

| | |
|---|---|
| Base effort (Person hours) [B] | 1000 |
| Total number of operating system [n] | 4 |
| Modification percentage per operating system [m] | 15 percent |
| Modification effort per additional operating system (person hours) | 150 |
| Total effort (person hours) = [B+(n-1)*m/100*B] = [B{1+(n-1)*m/100}] = Base effort + 150 hours each for 3 additional operating systems | 1000+3*150 = 1450 |

# Return on investment

Test automation entails a significant investment for any project. With mobile automation, this investment is even higher due to the various costs such as the device and additional tool plugin license costs involved. Hence, it is very important to be able to predict the economic viability of the venture to avoid losses later. This is the same as any investment decision that an organization is required to make before committing funds to any business proposition.

In this section, you will learn about this very important, but often overlooked, aspect of a mobile test automation project. You will learn how RoI is calculated, how to accurately analyze and predict the time it would take for a mobile automation project to become profitable, and you will learn the best practices to be adapted throughout the mobile test automation project lifecycle to maximize and speed up the realization of a positive RoI.

# RoI calculation

The return on investment for any automation project stems from the fact that the cost of execution of an automation test suite is negligible as compared to the manual execution cost. If the manual execution cost is repeatedly incurred, then the cost of the quality of the project keeps on increasing. Due to this fact, it is often the case that only regression test cases that are required to be repeatedly executed are selected for automation. In recent years, with the **behavior-driven development** (BDD), **DevOps**, **Shift Left**, and **Early automation** trends gaining popularity, organizations embark on test automation without performing any cost-benefit analysis, assuming that automation is always good and yields positive RoI. However, there is always a need to ensure that only the right candidates for automation are selected. This is where the RoI calculation helps.

To accurately calculate the RoI of any given project, first of all, you need to accurately determine the investment to be made for automation. The following are the factors that impact the investment cost of a mobile test automation project:

- **Cost of licenses** (CI): With mobile automation, apart from traditional tools, there is usually a need to invest in add-ons that are specifically developed to support mobile automation.

- **Cost of devices** (Cp): Either with real devices or with cloud-based device usage charges, mobile test automation costs are considerably increased. With real devices, apart from the device cost, there is an additional cost of maintaining infrastructure and the cost of network usage, such as GSM rentals and data usage charges.

- **Cost of development of automation suite (Cd)**: This cost includes the effort it would take to develop an automation suite including the one-time framework development cost.

- **Cost of Execution and Maintenance (Cem)**: This is the recurring cost of any automation project.

These costs are then compared against the manual test case execution cost to check when the break-even point is realized and whether it is within an acceptable time frame for the organization.

 The break-even point is the moment in a project execution when all the initial and ongoing investment is paid off, and from that moment onwards, all yield is considered positive.

In mathematical terms, the return on investment can be denoted by a simple formula, as follows:

$$Return\ on\ Investment = \frac{Net\ Benefits}{Net\ Investment} \times 100 = \frac{Cost\ of\ Benefits\ (Cr)}{Cost\ of\ Ivestment\ (Ci)} \times 100$$

The benefits denoted in the previous equation as **Cr** can be calculated as the actual cost saved by utilizing automation as against the cost it would have taken if the same testing was done manually, that is, the **cost of investment** denoted by **Ci** in the equation.

The cost of investment is generally a one-time investment, which is the sum of all costs borne while developing and maintaining the automation suite, as shown in the following formula:

$$Ci = \sum (Cl + Cd + Cp + Cem)$$

The formula to calculate the cost of benefits *Cr* is as follows:

$$Cr = (Cost\ of\ Manual\ Testing - Cost\ of\ Automated\ execution\ and\ Analysis)$$

For example, let's consider a mobile web application testing project for which there are four devices in scope. For each device, there is a need to test 150 test cases that require an effort of 20 person days (considering the usual productivity of a manual tester to be 7 to 8 test cases per day). So, in total for four devices, the cost of execution would be 80 person days. If we consider that a manual tester works for 8 hours a day with an hourly billing rate of $20.00 per hour, then the total cost of manual execution on all four devices is (8*80*20=) $ 12,800.00 per execution cycle.

For the same project, the cost of development of an automation suite with the user-agent-based automation using Selenium as an automation tool can be calculated as follows:

Considering that average test automation scripts productivity is 2 automation scripts per day, for an automation suite of 150 test scripts, the total effort for script development on one device is $150/2 = 75$ person days. Since the application is a web application, there is no additional effort consumed to adapt the scripts for other devices. Hence, the total effort remains 75 person days, which is $75*8 = 600$ person hours. Considering the hourly billing rate for an automation tester as $ 30.00 per hour, the total cost of automation suite development is $600*30 = 18,000.00$ dollars. For the execution setup and analysis of each device, if we consider an effort of four hours, then for four devices, the total effort for execution is $4*4 = 16$ hours. The cost of this automated execution effort is $16*30 = \$480.00$ per execution cycle.

So, using the formulas, RoI can be calculated as follows:

1. $Ci = \$ 18,000.00$.
2. $Cr = \$12,800.00 - \$480.00 = \$12,320.00$.
3. $RoI = Cr/Ci = 12320/18000*100 = 68.4$ percent per cycle of execution.

This means that with each cycle of execution, the automation suite returns 68.4 percent of the initial investment done to develop the automation suite.

We will see how to analyze this data in a graphical manner in the next section.

# The RoI analysis

Considering the example given earlier, we will analyze the return on investment for the project graphically using the following table:

| Release | Device | Manual cost | Automation cost |
|---|---|---|---|
| R1 | D1 | $ 3,200.00 | $ 18,480.00 |
| | D2 | $ 6,400.00 | $ 18,960.00 |
| | D3 | $ 9,600.00 | $ 19,440.00 |
| | D4 | $ 12,800.00 | $ 19,920.00 |
| R2 | D1 | $ 16,000.00 | $ 20,400.00 |
| | D2 | $ 19,200.00 | $ 20,880.00 |
| | D3 | $ 22,400.00 | $ 21,360.00 |
| | D4 | $ 25,600.00 | $ 21,840.00 |
| R3 | D1 | $ 28,800.00 | $ 22,320.00 |
| | D2 | $ 32,000.00 | $ 22,800.00 |
| | D3 | $ 35,200.00 | $ 23,280.00 |
| | D4 | $ 38,400.00 | $ 23,760.00 |

RoI Analysis Table

The highlighted row in the table denotes the break-even point. This is the iteration during the execution cycle, in which the total cost of automation becomes less than the total cost that would have been incurred had the execution been done manually. This can be plotted on a graph as follows:

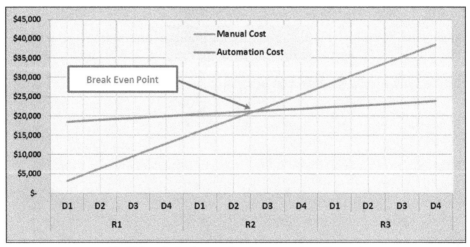

The RoI analysis of manual cost versus automation cost

So, we can see that as the project proceeds to the execution phase, the cumulative cost of automated execution does not grow at the same rate as that in the case of pure manual testing. Hence, after a few iterations, the overall cost of automation becomes less than that of manual testing. An alternative way of representing the RoI data is as follows, with an assumption that in every year, there are four releases of the application and the devices under test remain the same:

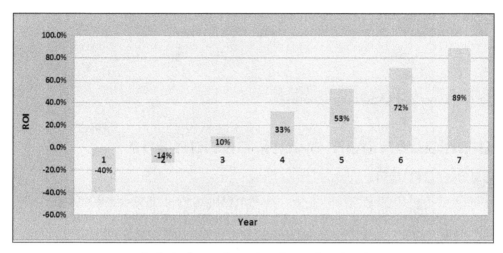

Analysis of percentage-wise return on investment

In this section, we have focused on the hard dollar RoI that the automation suite provides. However, it is also very important to realize that this mechanism does not yield all the benefits that test automation provides. There are plenty of intangible benefits such as reduced time to market, increased confidence to roll out newer releases in the development and design teams, consistency of results regardless of individual tester inputs, no risk of human errors, and better compliance with regulations and standards. These benefits are intangible benefits of automation and cannot be measured in dollar terms, but these are equally important while deciding to perform automation.

Till now, we have a clear understanding of all the planning stages of a mobile testing automation project. In the next section, we will look at the execution stages of a mobile testing automation project with a view toward introducing the best practices to maximize the return on investment of the project.

# Best practices to maximize the RoI

A good plan can only be as successful as the execution of it. Hence, while looking at the execution phase of a mobile automation project, we will look at the various stages and focus on how to deliver an efficient execution. In this section, you will learn about the execution phases because, in these stages, we will develop and execute the automation suite. These stages are:

- Automation framework design and development
- Test scripting
- Execution
- Extensibility and maintenance

# Automation framework design and development

A general rule of thumb to remember while working on a mobile automation project is that the return on investment for a project is as quick as the effectiveness of the automation framework to bring down the cost of automation. So, the RoI can be maximized by designing and developing an automation framework that helps to bring down the execution cycle time, reduces the script execution log analysis time, and improves the amount of test coverage that can be achieved. During the tool selection phase, a tool should be selected not just by keeping in mind the license cost, but proper weightage should be given to the ease of use and the scripting productivity that can be achieved with it. A mobile automation project is more lucrative than a traditional functional automation project because it allows the execution of the same set of scripts across multiple sets of devices, which means that the test suite development costs are more quickly adjusted against the manual execution costs.

An optimized test automation framework can be designed while keeping in view the following aspects:

- Coding and commenting guidelines are clearly defined for the support libraries and automation scripts, including business components, database connection, validation steps or check points, as well as loggers and reporter event steps.
- Application object naming conventions are clearly defined and enforced with a well-maintained object repository of the application. For device-specific objects, the framework should ideally enforce the name of the device within the object name.

- Integration with a version-control system is inbuilt with well-defined processes for library versioning.

- Documentation of the framework is well maintained so that future enhancements are easy to perform and manage.

- The framework should allow nontechnical users, the use of the framework in the form of keyword libraries that can be pulled together with specific test data to create customized test scripts.

- The test data management policy should be enforced to keep test data separate from the test script and the parameterization of test data should also be enforced. If there is need to maintain device-specific test data, then it should be kept in separate data sources to maintain the encapsulation of such data from test-case-specific test data.

- The framework should ensure that to update the test data, you should not require updated code, and this can be achieved only by updating test data input sources such as XML, Excel, CSV, or text files.

- The framework should allow a single-click execution trigger for all the scripts. This can be achieved by the use of a central driver script or with a common execution manager.

- The framework should not require any additional steps to launch devices and it should be programmatically handled.

- A graceful exit mechanism should be built at the framework level so that, in the case of any catastrophic failure during test suite execution, the execution would not go into an infinite loop or take an unreasonably long time to complete the execution.

- The framework should be integrated with any third-party tools or extensions required for mobile test automation.

# Test scripting

During the test scripting phase, the focus sometimes gets diluted and the teams just work on delivering the executable scripts rather than keeping in mind the guidelines laid down with the framework. This leads to an ad hoc test script creation and, especially with mobile automation, it may result into duplication of effort for different scripts for different devices and even for the same application functionality.

To avoid these pitfalls, the recommended approach is as follows:

- **Define device-wise reusable components**: All the components of the framework to be scripted should be made as reusable as possible. For example, a reporter function should accept the device ID set at the framework level during the execution as a variable and not use a hardcoded value. This makes the same component reusable throughout the automation suite rather than developing multiple components for various devices.

- **Business-process-based scripts**: During the scripting design phase, reusable business components must be identified and developed for those individually, while at the same time keeping in mind the inter-device and inter-operating system dependencies. For example, the login component of an application is a business process hence, it should have only one implementation throughout the project. However, the process to perform the login action might be slightly different across Android and iOS operating systems. Hence, for these two operating systems, device-specific handling should be done at the Login component level rather than having two separate business components as Login_Android and Login_iOS. This can be achieved by adding a conditional check in the Login component, so that, it self-detects the device under test and activates the required portions of the code.

- **Maximize the reusability of all supporting components**: Not only must the framework enforce the reuse of business logic and components, all the developed libraries for a project must also be shared across the organization/project team/product team. This is because in many cases, different teams across an organization may need to develop automation suites individually for the same application. The reusable utilities and components should be designed for the following types of elements and any device-specific steps, should be kept configurable, so that while calling the components, the device-related information may be provided as a variable:

    - **Actions to be performed**: Actions to be automated for different objects of the application should be kept configurable.

    - **Communicating systems**: The data exchange and flow between different internal systems of the framework and application or third-party systems should be handled in individual components.

    - **Database communication**: Database validation and check point validations should be made in a universal way for the application rather than keeping them aligned to specific test steps in business components. This provides the added flexibility of changing database sources as per different execution environments such as different environments used for the **Unit Testing (UT)**, **System Integration Testing (SIT)**, and **User Acceptance Testing (UAT)** phases.

- ° **Integration with other automation tools**: In some scenarios, the mobile automation framework might need to be integrated with a desktop- or web-based application that requires it to be integrated with some preexisting automation suite developed with another automation tool. For such cases, the components that trigger scripts developed in other automation tools should be developed in such a way that, with a single call, the respective scripts or reusable components of the third-party suite may be triggered.

- ° **Data retrieval components**: Retrieval of data from multiple input data files such as Excel, text, and XML files should be encapsulated within a single component, so that, during the scripting phase, the automation script developer team does not have to work with many components.

- ° **Execution schedulers**: Functionalities related to the process of invoking relevant scripts as per scheduler configuration should be developed as separate components or kept integrated with the execution manager component.

- ° **Mobile device communication**: Device-communication-related and data-transfer-related steps and validations should be developed as separate components. Especially with cloud-based tools, care must be taken to keep these components separate as well as configurable, so that they invoke the correct device as per data provided by the execution manager component.

- ° **Loggers**: Apart from the execution reports that are generated for end user consumption, scripts- and framework-level debugging information should be captured and logged in user-defined log files for retrospective analysis and debugging.

- ° **Error handlers**: Error handlers are used to handle known exceptions within a mobile environment, such as network outage, battery low reminders, and SMS popups, and other unknown errors should be developed to handle them gracefully and log the information in test results. This ensures that there is no need to build such exception handling mechanisms at the test script level.

○ **Test result creation components**: As per specific customer requirements, the execution reports should be generated on the completion of the test execution. These reports should ideally have all the device-specific information in a common header, so that each report can be specifically identified with an individual device, which would help reduce the defect recreation time. Also, it is recommended that the test result should also capture screen shots of the devices during the execution of the test, so that debugging and analysis of the test do not need manual intervention to replicate the defect.

In *Chapter 2, Designing Mobile Automation Frameworks*; we discussed the automation framework design specific to mobile automation. So, refer to it for practical implementation of these guidelines.

# Automation suite execution

During the test execution phase for a mobile test automation project, a lot of effort may go wasted if the framework design for the proper execution of the automation suite is not completed, or even when the execution plan is not efficiently made. During the test script execution phase, it is required that you repeatedly execute the same suite over multiple operating systems and device combinations. If this information is not easily configurable, then the test execution becomes a very tedious process even with automated scripts. Along with this, the test data sometimes needs to be configured in a device-specific manner, meaning that if this process is time consuming, then the automation suite execution is not efficient enough.

The following factors need to be considered for the efficient automation suite execution of a mobile test automation project:

- **Structure scripts with minimum human intervention required**: It is very important to structure scripts with minimal interdependencies so that scripts can be executed in an unattended manner even in the case of failures.

- **Provision for single click execution trigger**: With a single click or batch-file-based execution mechanism, the schedulers can be configured to trigger the execution within a specific time period. This saves the manual effort required to set up the execution.

- **Effective utilization of continuous integration**: Using continuous integration tools such as Hudson and Jenkins, automation suites can be automatically triggered after the deployment is completed for a mobile application. For a mobile web application, this is a very similar process to the traditional automation, apart from the fact that devices can be preset or preselected for the defined regression run. However, for native and hybrid applications, the latest application version installation is also required before the start of execution. For this, the automation framework should make a provision to uninstall a previous build of an application and replace it with the latest version, thereby negating the need of human intervention every time a new version of the application needs to be tested. Continuous integration of processes helps to boost productivity and generates high RoI.

- **Trend analysis facility**: The automation suite should allow you to capture script-wise execution trends for faster and more efficient execution analysis. With the historical data present for the past three or four runs, the execution analysis gets done much faster for a testing team because if a functionality is reported as failed on a particular device and is known to have failed on another device in the previous run, then this would most likely mean that both the failures have the same root cause.

# The execution manager

To effectively manage the execution setup and trigger the quick analysis, an efficient execution manager is paramount for the success of the execution cycles. The following is a sample execution manager for a mobile test automation project. In this execution manager, all the test execution setup information is provided in a simple way. The status flag is updated at runtime to denote the execution success or failure. The execution flag would be used in the driver script to determine which of the scripts from the test suite are required to be executed. In this example, only the scripts with the **Execution Flag** column that are marked as **True** will be executed:

| Test script ID | Device ID | Device operating system | Browser | Application type | Execution flag | Status (pass/fail) |
|---|---|---|---|---|---|---|
| TS_01 | D1 | Android | Chrome | Web | True | |
| TS_02 | D2 | iOS | Safari | Web | False | |
| TS_03 | D1 | Android | Opera Mini | Web | False | |
| TS_04 | D3 | Windows | Firefox | Hybrid | True | |

Sample Execution Manager for Mobile Test Automation Project

So, with this type of setup, the execution can be easily controlled by making minor updates to the Excel file outside the script code.

 Along with the device information that is kept configurable at the test-script level, a universal setting can also be made at the framework-level, so that all the scripts run on a single configuration of device, operating system version, browser, or application type. This provides dual flexibility during the execution setup.

# Extensibility and maintenance

During this phase, the set of scripts in the existing automation suite developed in the previous releases are maintained to support the modified and newly added functionalities, that is, the new enhancements in an application to support new features. The changes may be required due to any of the three reasons:

- Changes to the screen flows of an application
- New test data may be required to execute the same flow
- In most cases, the properties of the application objects are changed or even the objects themselves are changed

To accommodate these changes in the most effective and efficient ways, the given guidelines can be followed:

- If there are flow changes, then only the impacted business components should be modified. However, if the flow changes are specific to a device, then they should be accommodated within the business components that are already present in the test library. New business components for devices should not be made.

- Global data, such as device IDs, should not be updated while deleting the old data. Rather, new entries should be made and old entries should be marked obsolete for retrospective log maintenance.

- If there are extensive changes required to object properties for a particular device, then a shared object repository for only the newly added or modified device should be created.

- While making these modifications and additions, the standard process for script versioning should be followed for ease of traceability and for future maintenance.

# Some more best practices

Let's take a look at the following best practices:

- Sometimes, testers make the mistake of measuring complexities where the upper limits are not defined. For example, a complex test case is sometimes defined as having more than 30 steps. In this case, since the upper limit is not defined, all the test cases that have more than 30 steps are counted as a complex test case. This leads to incorrect estimation because without a clearly defined upper limit, even a test case with 150 steps would be considered as only one complex test case. So, it is very important to define the upper limits of test cases and then count the test cases in multiples of the complexity definition's upper limit. Thus, when the test case complexity is defined with an upper limit, such as having steps between 20 to 30, all the test cases that have more than 20, but fewer than 30 steps are considered as one complex automation script for estimation purposes. A test case with 150 steps would need to be considered equal to *(150/30=) 5* complex test scripts for effort estimation.

- To test the applications deployed in multiple languages, the execution can further be spread across devices, operating systems, and language combinations. A rule of thumb is that optimization can be done over any number of parameters, but once defined, all test cases that are part of the automated regression suite must be executed regularly and no further reduction in test script count is advisable.

# Summary

In this chapter, you learnt the basic principles that bring together all the technical aspects of a mobile test automation project. The ideas presented in this chapter will help any project manager and architect of an automation project to take informed decisions and develop designs that are suitable to their own projects. We discussed the factors that impact the RoI of a mobile automation project in detail, so that teams can take decision keeping in mind the optimization of the return on investment of the project. In the next chapter, you will learn about the practical implementations of these principles to understand how all these aspects can help to deliver a delightful customer experience to both the end users as well as the organizations that you would serve.

# 8

# Delivering Customer Delight

Having learnt all the fundamental and technical implementations of various mobile automation methodologies and approaches in the previous chapters, in this chapter, we will learn about utilizing these concepts in real-world scenarios. We will learn about various aspects such as the importance of test automation with respect to improving RoI of a project and how it is more logical to perform automation in a mobile testing project than in regular functional testing projects. We will also explore other important aspects such as the use of emerging technologies in functional automation, such as continuous integration in mobile test automation projects. Apart from these, we will look at the architecture stack and various logical components of an ideal mobile test automation framework to bring together all that we have learnt in the previous chapters.

## Customer delight – internal and external

Although **customer delight** is a self-explanatory term, it is often a chimera that project delivery teams run after because they really do not have a clear understanding of what customer delight signifies, and, more importantly, how to achieve it. This is partly because of the fact that it is more of a *feeling* than a measurable entity. However, in the mobile world, the star rating system is an instant feedback mechanism for a mobile application and provides a way to measure the level of external customer satisfaction for a mobile application.

So, knowing who your customers are is the first step toward understanding what your customer needs, and, more importantly, what their wants are, and only then, will you be able to surpass those to achieve what is termed as "customer delight". There are two sets of customers that are required to be catered by any mobile testing automation projects. The final focus is always on the end user of a mobile application, who is the external stakeholder or customer. Their level of satisfaction toward an application is the ultimate goal for any project and therefore, for the automation team as well.

With the ease of communication that a mobile platform provides, not only are external customers very keen and quick to grant five-star ratings to reflect their delight with the overall mobile application experience, but they are even quicker to voice their disapproval on social forums, such as Facebook and Twitter, or the marketplaces, such as Google Play Store or iTunes. The negative reviews can in turn impact the level of satisfaction of the internal stakeholders as well, since negative public reviews and ratings can jeopardize the credibility of an organization instantly and often irrevocably harm the public's perception of them.

So, apart from the obvious external customers, with mobile testing automation, there is also an equally important need to satisfy the internal stakeholders, who can be the business owners of an application in the organization, the development team of the application, or even the functional testers of an application. These functional testers would want to use the automation suite to perform repeated tests of an application with every new build, using the automation solution, thereby focusing their manual effort on testing new functionalities. Hence, not only does the automation solution need to ensure a great application is produced, it needs to be done efficiently, in the simplest form possible, without needing any manual intervention, and in the most cost-effective way possible in order to generate internal customer delight.

From the external customer viewpoint, it is very important not only to deliver a defect-free application but also a positive user experience with the mobile application. So, with automation testing, the goal should be to test the responsiveness of the application and to track instances when customer data may be potentially lost due to incoming interrupts during a flow, and also to ensure that all of the desired requirements are in compliance with the functional requirements.

Consequently, with mobile automation, it is not enough that we are able to identify defects, but we must also be able to effectively measure the overall quality of the application. The automation solution results should provide a way to gauge the application's performance and user experience aspects as well. So, cases such as an application crash, which are not considered in a normal functional testing project as part of the scope of an automation solution, have to be specially handled in mobile automation projects apart from the normal **Nonfunctional Testing (NFT)**, such as performance and load testing, which is carried out for a mobile application.

In later sections, we will take a look at a set of sample mobile testing projects and how test automation can be utilized to deliver internal customer delight, apart from the obvious reduction in production defects, which is the usual function of an automation solution. These examples are taken from real-time projects and the solutions are discussed to understand not only the automation scope, but the overall project delivery as well, with specific focus on functional testing and test automation processes that results in generating internal customer delight.

Let's now look at a few important aspects such as the architecture of an ideal mobile test automation framework and the use of the **behavior-driven development** (BDD) methodology along with use of **continuous integration** (CI) tools in a mobile test automation project, which provide us with the capability to continuously monitor the application code quality and adhere to the acceptance criteria of the **Application Under Test** (AUT).

# An ideal mobile test automation framework

Internal customer delight naturally stems from a realization that the solutions delivered are future-proof and can be incrementally used as the need arises. In this section, you will understand the architectural stack of an ideal mobile test automation framework. The implementation principles and sample codes have already been discussed in the previous chapters, and in this section, we will focus on understanding the architecture of such a framework and the advantages that the implementation of such a framework provides. The architecture for an ideal mobile test automation framework is shown in the following diagram.

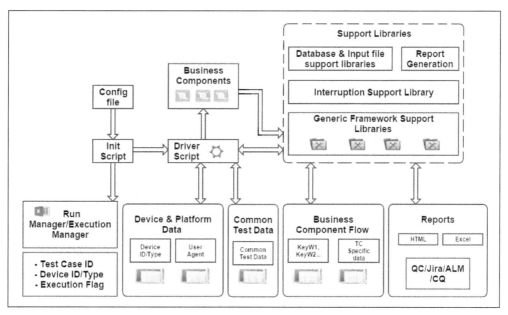

Architecture of an ideal mobile test automation framework

This diagram depicts all the logical components individually; however, from a coding perspective, two or more components may be combined with each other.

The various components of the framework and their respective functions are explained as follows:

1. **The configuration file**: The configuration file can be a simple XML or TXT or any other file format with global information of the AUT, execution environment, application, and DUTs. Keeping this information in a separate file allows independent updates by anyone (a manual testing team or business users — internal customers). For example, if all the test cases for an application need to be tested with a Wi-Fi connection instead of the LTE network, then this control can be put into the configuration file so that this change happens for the entire set of scripts and the execution can be triggered with a small change.

   ° **The initialization script**: This script should read the data from the `Execution Manager` file and the global framework-level data present in the configuration file and provide it to the main driver script of the framework. This is a logically independent section of the framework; however, from an implementation point of view, this can be integrated with either the configuration file (in the form of a macro of an Excel sheet) or with the driver script (as part of the *@Before Test*, *@Before Run*, or *Setup* sections that are reset in the *@After Test*, *@After Run*, or *Tear Down* section). It is basically the entry point of the script execution. This initialization script can also be in the form of a simple batch job that instructs the automation tool, such as Selenium, to set up the grid and trigger execution.

   ° **The execution manager or run manager**: It should contain the test script data to allow various internal customer teams to set up the test suite execution as per their own needs with the required device IDs, device types, and other relevant data.

   ° **DUT data**: This table should contain the relevant device under test (DUT) data. For a framework that uses user agent, DUT data should contain the user agent string data along with the operating system and device type information, and for cloud connections, it should have the device connection information to be used for the cloud tool setup.

- ○ **Common test data**: Common test data that does not change for the entire test suite, such as the web application test environment URL, should be kept in a tabular format in this file. Each entry can be given a representative ID and in the business component code, rather than the actual test data, it can be referenced with that ID. This technique is called the relative parameterization of the test data and helps to update the test data appropriately without major changes in the test data input file.

- ○ **Business processes**: These are the business component flows that are stored in tables with specific data as per the test cases. These should ideally be self-contained and can be reused across test cases. Using the business components, application experts—that are the internal stakeholders—can create their own flows by calling them in sequence and with specific application test data.

- ○ **Business component code**: In case a flow changes, then updating the code automatically updates all the business processes and test cases that use the code.

- ○ **The object repository**: This should be kept as a logically independent entity in the framework, so that the process updating the scripts to make changes in the application is streamlined in a better way. For tools such as Selenium that do not explicitly support object repositories, this logical component can be easily designed by creating class files that contain only the application object information.

- ○ **Support libraries**: All generic support libraries should be grouped together so that an architect doesn't have to change the automation script code in order to make the framework-level changes to the automation suite. These files should contain the following logically independent blocks:

  - ○ Report generation
  - ○ Interruption support library
  - ○ Database and input file support libraries
  - ○ Generic framework support library such as integration with QC, JIRA, or error handling and recovery

The use of such an automation framework provides the following advantages that ultimately lead to a smooth delivery and, in turn, generate a delightful customer experience. For internal customers, this framework ensures a simple-to-use solution, and for external customers, it helps to provide a better production quality by allowing faster execution cycles over multiple devices and platforms simultaneously, as explained in the following section:

- This framework architecture can support the automation solution without any modifications required for different automation methodologies and tools.

- With the same file stack and interactions, internal customer teams would not have to work with different frameworks and approaches, resulting in reduced **total cost of ownership** (**TCO**) and better RoI.

- With a single framework for all approaches and tools, the learning curve required for people to start delivering automation scripts will be reduced significantly.

- Tool- and DUT-specific data is externalized from the test data, with the only difference being separate language-specific implementations for specific tools.

- Such an architecture allows the same script to be reused for different mobile handsets instead of having multiple automation scripts for each manual test case.

- The independent interruption testing component in the framework library makes it easy for automation teams to be integrated with business components.

- The same script can be executed across multiple platforms with a single configuration change using the configuration file and thus, there are no code-level changes required for the execution setup.

- This structure provides flexibility to execute test scripts individually on different mobile operating systems and devices as well as scheduling the entire suite on the same DUT.

- Application data is mandatorily stored externally in the scripts of data tables, thus ensuring data parameterization.

- **Support for all mobile automation approaches**: When user agent, emulators, real devices, clouds are based on the same framework, this means that internal customers do not have to learn different techniques and solutions.

- **This structure can support all mobile automation tools**: QTP, UFT, Selenium, Ranorex, SilkTest, Perfecto Mobile, Mobile Labs, and others includes tool and language-specific libraries, but since the structure remains the same it is easier to integrate different approaches within the same solution. For example, a set of scripts can work with only user-agent-based automation and another set can work with real devices without any modifications required to the framework.

- A singular framework for all approaches can lower the time to market because the changes done would be simultaneously applicable to all scripts within the scope of an automation suite, thereby leading to faster delivery.

# Mobile automation for agile projects

Given the agility that a mobile application development demands, any application team would be most interested in techniques that allow shorter release cycles and faster validation of newly coded features. This is why most organizations now follow an agile delivery model. In this section, we will look at certain methodologies and concepts that are useful for agile projects.

# DevOps and shift left

**DevOps** is the short form of **Development and Operations** and indicates the synergy that needs to be achieved between all concerned teams for a project delivery. It combines the lean and agile schools of thought and aims to enable testing of applications that are as close in approximation to the production environment as possible. Most organizations working with mobile applications would want to roll out updates as quickly as possible, and the DevOps methodology allows them to achieve this in a streamlined manner by ensuring that the tests and development both start on a production-like environment as early in the lifecycle of the process as possible. Also the operations teams (that support production environments and servers) are therefore able to spot potential bottlenecks early.

With the use of the DevOps methodology, it is also possible to perform tests earlier in the lifecycle because it mandates that the testing and development is done on a system that behaves closely to the production environment. Hence, with the use of drivers and stubs to replicate upstream and downstream systems, the testing basically shifts toward left in the lifecycle. The use of a system to virtualize the web services feeding a downstream system is a practice very commonly associated practice with the shiftleft approach. However, it is not limited to only that.

Any method with which testing can be started early in the lifecycle can be termed as **shift left**. So, the following are the principles that are required to be achieved in any project:

- Continuous integration
- Continuous delivery
- Continuous testing
- Continuous monitoring and feedback

# Behavior-driven development

**Behavior-driven development (BDD)** is a methodology of software development and testing that focuses on the desired behavior of the application. This is an evolutionary outcome of the **test-driven development (TDD)** approach of application development. It mandates high automation of the quality assurance aspect and focuses on the creation of automation scripts that can be used to verify the specifications against the acceptable behavior of the application.

As a methodology, BDD can work with any language or tool of choice, but it is most commonly used with the **Cucumber** framework integrated with tools such as **Web Application Testing in Ruby (WATiR)** and Selenium. Cucumber specifies the software behavior tests in the language of the business in a specified syntax of grammar rules called **Gherkin**. Gherkin ensures that the acceptance criteria are written in terms of scenarios and implemented as classes in the following format:

- Given [initial context]
- When [event occurs]
- Then [ensure some outcomes]

In the following example, the `credit card validation` feature of an application is shown. The feature also has two negative scenarios—first, with the credit card number being too short and second, where the number is too long.

```
Feature: Credit card validation.
Credit card numbers must be exactly 16 characters.

Scenario: Credit card number is too short
  Given I use the native keyboard to enter "123456" into text
  field number 1
  And I touch the "Validate" button
  Then I see the text "Credit card number is too short."

Scenario: Credit card number is too long
```

```
Given I try to validate a credit card number that is 20
characters long
Then I should see the error message "Credit card number is too
long."
```

There can be any tool-specific bindings for these feature files. For example, the Ruby code behind the feature file for the second scenario is as follows:

```
Given(/^I try to validate a credit card number that is (\d+)
characters long$/) do |20|
  touch("textField marked:'CreditCardNumberField'")
  wait_for_keyboard
  keyboard_enter_text("7" * number_of_digits.to_i)
  touch("button marked:'Validate'")
end

Then(/^I should see the error message "(.*?)"$/) do
|error_message|
  text_view = querjh bgy("textView marked:'ErrorMessagesField'
  {text CONTAINS 'Invalid Card Number'}")
  raise "The error message '#{error_message}' is not visible in
  the View." unless text_view.any?
end
```

 For more information on BDD, read the very informative and interesting blog by Dan North, the creator of BDD, at http://dannorth.net/introducing-bdd/.

With mobile testing automation frameworks, Cucumber can be integrated with Selenium to support the BDD methodology. Also, some tools such as **Calabash** by **Xamarin** support cross-platform acceptance tests in Cucumber, which can be used to run acceptance tests across iOS and Android. Xamarin also provides a test cloud solution with ready-made integration for these tools.

# Continuous integration for mobile automation

The test harness for an automation suite must also facilitate continuous, repeated, and automatic triggering of a test execution.

**Continuous integration** (CI) refers to the practice of periodically integrating all newly developed and modified components within an application code base and then running automated tests to determine the code stability and their adherence to the requirements. Running automated scripts in a completely handsoff and unattended mode ensures that integration bugs in the code are identified at earlier stages of testing and ensures that the root cause is tracked down. Some commonly used tools for CI are **Hudson**, **Jenkins**, **Team Foundation Server** (**TFS**) and **TeamCity**. With the use of CI tools, the automation suite can also be executed in parallel on multiple mobile devices simultaneously and the testing turnaround time can be significantly reduced. Now, you will learn about the integration of mobile automation solutions with Jenkins.

Jenkins can be integrated with HP UFT with the use of a plugin. The plugin ID is called **hp-application-automation-tools-plugin**, which can be downloaded from the HP support site. This plugin allows you to trigger an HP test as a build step and provides the results within the Jenkin's UI, similar to any JUnit test, which can later be used to notify functional testers and developer teams about the failures.

This plugin supports HP **Unified Functional Testing** (**UFT**), and the HP **Application Lifecycle Management** (**ALM**) tool can be integrated with the QC plugin. With Selenium, the Selenium plugin provided with Jenkins may be use, which sets up a Selenium grid that can then be used for execution. The grid can consist of mobile devices as nodes, meaning that the execution is spread over numerous devices in parallel.

Apart from this technique, with the use of the ideal automation framework explained in the previous section, either of the `Init` or `Driver` files can be configured as a post-build deployment job on the Jenkins CI server. So, this kind of setup can be leveraged with cloud devices, real devices, emulators, and user agents. This CI setup is depicted in the following diagram that delivers an end-to-end continuous integrated execution platform:

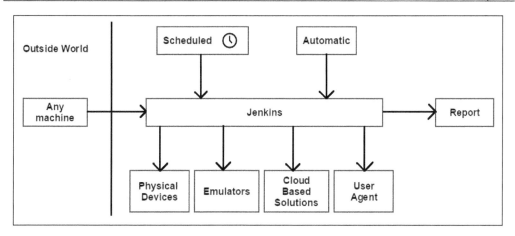

Jenkins can be used to automatically execute over multiple systems and with a variety of triggers. Calabash also provides out-of-the-box integration with all CI solutions such as TFS, Jenkins/Hudson, and TeamCity, with customizable post-build commands.

# Some sample mobile projects

In this section, we will take a look at a set of sample mobile testing projects and the solutions that resulted in customer delight.

# Project 1 – an insurance provider's web application

The AUT is an insurance provider's self-service web application developed in HTML5, using which, the consumers of the insurance company can register their insurance policies and then get related details of the insurance, such as asset allocation details, viewing a renewal premium payment calendar, and the claim process. The customers are also allowed to view premium receipts of previous payments made. The company has decided to only support Android 4.1 or higher and iOS 7 or higher version mobile phones, considering their target consumer base. For tablet devices, the customers will be presented with the normal web application and not mobile-specific web views.

# Testing requirements

Given that there are just view-only features, the testing team has decided to not include any specific test cases for the **American Disability Act (ADA)** testing and focus primarily on the functional testing, GUI testing, and basic interrupt testing.

## Optimization

Since the focus is only on Android and iOS's latest stable versions with maximum market share, the testing team recommended to use only four devices in total with a device and browser mix, as follows:

| Device OS | Browser |
|-----------|---------|
| iOS | Safari |
| Android | Opera mobile |
| iOS | Google Chrome |
| Android | Native/Chrome |

The preceding table provides a coverage of four different browsers on the two operating systems. This type of device selection provides maximum amount of coverage without needing all four browsers to be tested on the two operating systems, that is, a total of eight combinations, and works with only four combinations instead.

## The solution

With the primary focus on functional testing, a total of 50 test cases are estimated, averaging about six test cases per screen. Along with the functional test cases, about 30 GUI test cases are written, and for interrupt testing, another 10 test cases are designed. Since both Android and iOS are modern operating systems, the test automation solution can work. No specific ADA or UX test cases were estimated or required to be developed.

For the automation solution, Selenium with Java-based implementation is utilized with a hybrid automation framework. Since there was only a set of two devices required with two browsers set up on each, the team decided to use real devices. To control the execution across devices, a configurable execution manager with separate entries for all operating system versions was utilized. This ensured that in every execution, the scripts were always executed over all the targeted devices.

# Project 2 – automobile manufacturer's car dashboard and health-check application

This is a mobile hybrid application that supports iOS and Android to display and analyze various automobile's various performance data and presents the data to help diagnosis and health checkup of the automobile. The application connects with the car's dashboard-based display using a Bluetooth connection to collect the data. The application is also used to provide periodic updates to the driver to schedule maintenance.

## Testing requirements

The testing is required for iOS- and Android-based devices. The application is required to support two sets of end users, that is, the mechanics who will use the diagnostic data and the automobile owner, so that they get notifications about the required maintenance so the test cases are used.

## Optimization

Since it is not practical to arrange an automobile test during the early stages of testing, the testing team recommended to use dummy test data that replicates the inputs from an automobile. For the devices, testing was recommended to be done on emulators rather than real devices.

The OS versions to be covered are as follows:

- iOS 7
- Android 4.4
- iOS 6
- Android 4.3

## The solution

A basic set of 50 test cases were developed for functional and GUI testing. There was no requirement to perform interruption testing. For the automation solution, Appium with a Java-based automation on emulators and simulators was developed using a hybrid automation framework. Since the test data was required to be handled separately, the device setup module was kept separate, requiring no additional setup for device changes.

# Project 3 – using a web and hybrid application to enable BYOD for a secure banking application

The AUT is a portal that allows access via a web interface as well as a hybrid application developed using HTML5 with a responsive web design. The AUT has differing user interfaces for tablet and phone devices. The customer has decided to perform the project in phases, described as follows:

- **Wave 1**: Enabling **Bring Your Own Devices (BYOD)** for its employees who work as account managers for high-net-worth company. In this phase, the employees will be allowed to use their own tablet devices, such as iPads and Samsung tablets. The web application would be of the same view as the existing application that the account managers/advisors are using, without any changes to the user interface. For Windows and Blackberry, there would not be a stripped-down version of the web application created with only basic controls such as links and text boxes. This classic version of the website will have limited UI features; however, it will support all application functions as the enhanced version of the application with the use of simple control changes, such as links instead of buttons, which are easier to work on a classic view device.

- **Wave 2**: In this phase, a hybrid application will be created to support the account managers with the same set of functionality as the web application, with minor enhancements to the UI for Android and iOS phones. As per the user access level, the application would allow different flows.

- **Wave 3**: After enabling the access via mobile phones, the hybrid application will also support the tablet devices and the HTML5 code will be tested for the support provided to tablet devices. This is going to be a testing-heavy wave, as the responsive design is expected to cater to the tablet views automatically without needing major code changes.

## Testing requirements

Since the project is going to be delivered in multiple phases, there is a need to develop a comprehensive regression testing solution, which can support the changes to the application in multiple releases. The application has different views for enhanced and classic devices, so the automation solution needs to be able to work with both the versions of the application as well as be able to support both the web application and the hybrid application that is going to be developed during the later phase of the project.

# Optimization

Since the requirements of devices are different with incremental releases, the testing can be planned and performed on all four operating systems in question—Android, iOS, Windows, and Blackberry. Since in the first wave the existing regression automation suite can be reused, the automation solution should be able to accommodate it. Along with the reuse of the existing automation suite, it would be useful to implement a long-term mobile manual and automation testing solution, which can be used over a period of time.

# The solution

A private cloud solution based on Mobile Labs was implemented in the program, which provided facilities to upgrade the devices periodically without any significant cost. With the cloud setup, it was feasible to automate GUI as well as interrupt testing scenarios. Although it seemed like the private cloud setup was the most costly option, it ensured that the devices are well maintained and the applications were always securely accessed. Perfecto Cloud was set up within the premises of the customer to set up a private lab that supported all the required devices. Over the various phases of the project, the devices were regularly kept updated and used by teams spread across different geographies. For the first phase, the automation solution relied on the utilization of the user agent approach with the existing automation suite, which was developed using Selenium. By making minor changes to the login component, the user agent component was integrated and executed. This yielded instant RoI as the cost to develop the framework and suite was negligible.

With the cloud tool, Selenium was used with the **Selenium Mobile Driver** provided by Perfecto to create an automation suite that could support all the devices in question.

# Best practices

Let's take a look at the following best practices:

- **Beta testing enablement with test automation**: Within four weeks of its launch, a typical iOS user is upgraded to the latest operating system version. So, the testing must not only happen when the application release happens, but it is equally, and in some cases, more, important to, test the application with the upgraded mobile OS.

- **Gathering code coverage statistics**: The use of code coverage statistics allows production of handy indicators for development tools. There are various tools such as **Emma** for Android and **Xcode** for iOS, where internal configurations can be done to gather this information. This is a very important information that can help to add or modify test scenarios to increase code coverage provided by any automation suite.

- **Enhanced reporting for mobile**: For mobile automation runs, it is important to capture the global information such as the device name, device type, operating system version, AUT version, and any other relevant data that is common to the entire test case, so that the manual analysis of the run is more efficient.

- **Execution of trend analysis**: Creating trend reports of executions based on previous runs is also a very useful technique that helps to reduce the manual analysis time. If it can be spotted that a script has failed in the previous run, then the analysis would not be required unless the associated bug is reported as fixed.

- **Unattended execution and reduction in false failures**: To make sure that there are no false failures during the run due to reasons, such as the device going offline, network fluctuations, and application slowness (which are otherwise measured in nonfunctional testing only), there can be a mechanism developed that retriggers failed test cases a given number of times before reporting them as failed.

- **Functional page response time capture**: Tools such as UFT and QTP provide the **Timer** feature, which can be utilized to measure the elapsed time during the page loads for an application. Although this information is not as accurate as commercial performance testing tools, it is still an indicator of the application response time that an average user experiences with an application. Such information can indicate any performance bottlenecks early in the lifecycle of an application and also add a check during the regression testing phase, so that such issues are always detected before they reach the production stage.

# Summary

In this book, we learnt about the technology, methodologies, and approaches required to perform mobile testing automation. The chapters are provided in as comprehensive manner as possible. However, given the vastness as well as the still emerging character of the mobile field, and given that the support for upcoming technologies, such as **Near Field Communication** (**NFC**), wearable devices, and mobile contactless payments with technologies such as **Apple Pay** or **iBeacon** is still very limited with automation tools, it is not possible to cover each and every combination of a project upfront. With the examples of real projects presented in this chapter, we hope that the usually hidden aspects of mobile testing automation projects are clearer to you, and that you will be able to identify the most suitable approach to devising well-suited testing strategies that are aimed at delivering customer delight.

# Index

## Thank you for buying
# Mastering Mobile Test Automation

# About Packt Publishing

Packt, pronounced 'packed', published its first book, *Mastering phpMyAdmin for Effective MySQL Management*, in April 2004, and subsequently continued to specialize in publishing highly focused books on specific technologies and solutions.

Our books and publications share the experiences of your fellow IT professionals in adapting and customizing today's systems, applications, and frameworks. Our solution-based books give you the knowledge and power to customize the software and technologies you're using to get the job done. Packt books are more specific and less general than the IT books you have seen in the past. Our unique business model allows us to bring you more focused information, giving you more of what you need to know, and less of what you don't.

Packt is a modern yet unique publishing company that focuses on producing quality, cutting-edge books for communities of developers, administrators, and newbies alike. For more information, please visit our website at www.packtpub.com.

# About Packt Open Source

In 2010, Packt launched two new brands, Packt Open Source and Packt Enterprise, in order to continue its focus on specialization. This book is part of the Packt Open Source brand, home to books published on software built around open source licenses, and offering information to anybody from advanced developers to budding web designers. The Open Source brand also runs Packt's Open Source Royalty Scheme, by which Packt gives a royalty to each open source project about whose software a book is sold.

# Writing for Packt

We welcome all inquiries from people who are interested in authoring. Book proposals should be sent to author@packtpub.com. If your book idea is still at an early stage and you would like to discuss it first before writing a formal book proposal, then please contact us; one of our commissioning editors will get in touch with you.

We're not just looking for published authors; if you have strong technical skills but no writing experience, our experienced editors can help you develop a writing career, or simply get some additional reward for your expertise.

## Learning Software Testing with Test Studio

ISBN: 978-1-84968-890-1          Paperback: 376 pages

Embark on the exciting journey of test automation, execution, and reporting in Test Studio with this practical tutorial

1. Learn to use Test Studio to design and automate tests valued with their functionality and maintainability.

2. Run manual and automated test suites and view reports on them.

3. Filled with practical examples, snapshots and Test Studio hints to automate and substitute throwaway tests with long term frameworks.

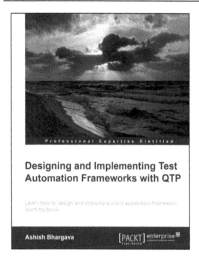

## Designing and Implementing Test Automation Frameworks with QTP

ISBN: 978-1-78217-102-7          Paperback: 160 pages

Learn how to design and implement a test automation framework block by block

1. A simple and easy demonstration of the important concepts will enable you to translate abstract ideas into practice.

2. Each chapter begins with an outline and a brief statement of content to help the reader establish perspective.

3. An alternative approach to developing generic components for test automation.

Please check **www.PacktPub.com** for information on our titles

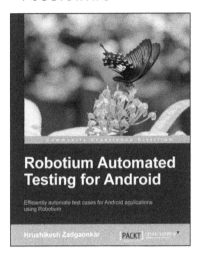

## Robotium Automated Testing for Android

ISBN: 978-1-78216-801-0          Paperback: 94 pages

Efficiently automate test cases for Android applications using Robotium

1. Integrate Robotium with Maven to perform test case execution during build.

2. Learn different steps to connect to a remote client from an android using Robotium.

3. Understand the benefits of Robotium over other test frameworks.

## Robot Framework Test Automation

ISBN: 978-1-78328-303-3          Paperback: 98 pages

Create test suites and automated acceptance tests from scratch

1. Create a Robot Framework test file and a test suite.

2. Identify and differentiate between different test case writing styles.

3. Full of easy- to- follow steps, to get you started with Robot Framework.

Please check **www.PacktPub.com** for information on our titles

www.ingramcontent.com/pod-product-compliance
Lightning Source LLC
Chambersburg PA
CBHW060530060326
40690CB00017B/3443